GENOMICS AND PERSONALIZED MEDICINE

WHAT EVERYONE NEEDS TO KNOW®

GENOMICS AND PERSONALIZED MEDICINE

WHAT EVERYONE NEEDS TO KNOW®

MICHAEL SNYDER

OXFORD
UNIVERSITY PRESS

OXFORD
UNIVERSITY PRESS

Oxford University Press is a department of the University of Oxford. It furthers
the University's objective of excellence in research, scholarship, and education
by publishing worldwide.Oxford is a registered trade mark of Oxford University
Press in the UK and certain other countries.

"What Everyone Needs to Know" is a registered trademark of
Oxford University Press.

Published in the United States of America by Oxford University Press
198 Madison Avenue, New York, NY 10016, United States of America.

Library of Congress Cataloging-in-Publication Data
CIP data on file with Library of Congress
ISBN 978–0–19–023477–5 (hbk.)
ISBN 978–0–19–023476-8 (pbk.)

1 3 5 7 9 8 6 4 2
Printed by Sheridan, USA

CONTENTS

ACKNOWLEDGMENTS   XI
INTRODUCTION   XIII

1 Personalized Medicine   1

What is personalized medicine?   *1*
What personal factors impact our health?   *2*

2 Genome Fundamentals   5

What is DNA?   *5*
What is a genome?   *7*
What does the genome do?   *9*
How does one person's genome differ from another person's genome? *9*
How do genomes of men and women differ?   *12*
How is the genome decoded?   *13*

3 An Introduction to Cancer Genetics   21

What is cancer and how does it arise?   *21*
How do the BRCA1 and BRCA2 genes cause cancer?   *24*

What are examples of other genes implicated in cancer? 25

How does genetic information help us treat cancer? 26

4 Genomics and Cancer Treatment 31

What has been learned from genome sequencing of cancer? 31

How can genome sequencing advance cancer treatment? 32

If I have cancer should I get my tumor genome sequenced? 37

Why do anticancer drugs fail and how might genomic approaches help address this issue? 39

Can genetics and genomics help detect early cancer and monitor treatment effectiveness? 41

A new approach: What is immunotherapy? 42

How can genomics be used to harness the patient's own immune system to fight cancer? 44

5 Solving Mystery Diseases 47

What is a mystery genetic disease? 47

How many Mendelian diseases are there? 49

How are genes responsible for genetic disorders identified? 50

How useful are genomic approaches to solving mystery genetic diseases? 53

Why can't most Mendelian diseases be solved? 55

6 Complex Genetic Diseases 59

What is a complex genetic disease? 59

How does complex genetics affect neurological diseases? 61

How does complex genetics affect metabolic diseases? 62

Can some diseases be both monogenic and complex? 64

7 Pharmacogenomics   67

*How can your genome directly help guide drug treatments
for treating disease?*   67

What are other ways your DNA can guide drug treatments?   67

Are there sex differences in drug effects?   69

8 Genomics for the Healthy Person   71

How can getting your genome sequenced improve your health?   71

Can genome sequencing affect the drugs someone takes?   76

*Can genetic testing be used to predict sports performance
and injuries?*   78

Will sequencing my genome affect my children and my relatives?   79

9 Prenatal Testing   81

How are genome sequencing technologies changing prenatal testing? 81

*Can genome sequencing be used to identify other mutations,
beyond chromosomal abnormalities, that might cause disease?*   83

*Can genetic testing be useful for choosing healthier embryos
and producing designer babies?*   84

10 Effects of the Environment on the Genome
and Epigenetics   87

How does the environment contribute to health?   87

When did people first begin to study environmental effects?   87

When do environmental effects begin?   88

Can environmental factors directly impact the genome?   89

What is epigenetics?   89

*What are some examples of environmental effects on physiology
that are mediated through epigenetics?*   91

What is the role of epigenetics in disease?   92

How will increased understanding of epigenetics impact health care?   92

11 Other 'Omes   95

What other 'omes are useful for medicine?   95

How can the transcriptome and proteome be useful?   96

How can the metabolome be useful?   97

How deeply can a person be analyzed?   99

12 The Personal Microbiome   101

What is the microbiome?   101

How is the microbiome studied?   102

How does the microbiome affect health?   104

How does diet affect the microbiome?   106

Can the microbiome affect other aspects of our lives?   106

Can the microbiome be altered to improve human health?   107

13 Your Immune System and Infectious Diseases   109

How does your immune system protect you?   109

How does the immune system vary among people and affect
our health?   111

How can we analyze infectious diseases?   112

14 Aging and Health   115

Are there genetic factors underlying longevity?   115

Are there environmental factors that affect aging and longevity?   117

Does epigenetics control aging?   118

Will we someday be able to control our aging?   119

15 Wearable Health Devices   121

*What other types of personal health information
can be readily collected?*   *121*

*How will this information be made available to, and used by,
the individual?*   *124*

16 Big Data and Medicine   125

How much data can be gathered about a single person?   *125*

How much data can be gathered about a group of people?   *126*

How can a large database assist in medical care?   *127*

How can Big Data guide lifestyle decisions?   *130*

What are the opportunities for industry in Big Data Medicine?   *130*

17 Delivery of Genomic Information   133

Who controls your genomic and other health information?   *133*

Who will deliver genomic information to you?   *134*

What is the role of the physician?   *135*

What are the implications of direct-to-consumer genomic testing?   *136*

18 Ethics   139

Can your genetic information be used against you?   *139*

*What are the concerns surrounding routine (or even mandated)
genetic screening?*   *140*

What is a possible solution?   *141*

19 Education   143

Can we educate people to understand genomic information?   *143*

How do we educate physicians to understand genomic information?   *144*

Who else should we educate? What is the role of healthcare providers, insurers, and policy makers?   *144*

20 Privacy   147

Can people be identified solely from their genome sequence?   *147*

Will having my genome sequenced affect my family members?   *148*

21 Paying for Personalized Medicine   151

Who pays for genome sequencing in treating disease?   *151*

Who pays for genome sequencing in preventive medicine?   *153*

Will genome sequencing make health care cheaper?   *154*

Will people act on genomic information?   *155*

22 The Future of Personalized Medicine   157

What other technologies will be prevalent in the personalized medicine space?   *157*

What will the future look like?   *158*

INDEX   161

ACKNOWLEDGMENTS

I thank Christine Costigan, Barbara Dunn, Yiing Lin, Shin Lin, Jon Bernstein, Ami Bhatt, Anne Brunet, Josh Gruber, Rajini Haraksingh, Fereshteh Jahanbiani, Chao Jiang, Stuart Kim, Liang Liang, Jason Reuter, Andrew Roos, Kun-Hsing Yu, Michael Wilsey, and Wenyu Zhou for many helpful comments on the book, and my laboratory for many inspiring ideas. Our research is supported by the National Instititues of Health.

INTRODUCTION

Imagine a world in which you can input your age, lifestyle, and genetic information (genome) into an app and obtain personalized recommendations about the food you should eat or avoid and behaviors you should modify to help maintain your health. Moreover, imagine that when you are sick, your physician inputs the same information to determine your customized treatment plan. That world is not 50 years into the future; in many respects it is beginning to unfold now. This book will reveal the essentials of personalized medicine (sometimes referred to as precision or individualized medicine), what is it, how will it affect you when you are healthy and when you are sick, and how personalized medicine will be implemented based on your DNA and other information.

1

PERSONALIZED MEDICINE

What is personalized medicine?

The practice of medicine has always been personal. Doctors use extensive personal information about a patient—including medical history, physical exam, vital signs, family history, laboratory measures, and imaging tests—to determine a patient's risk for certain diseases and to make diagnoses. So what makes medicine more personalized now? The answer is that we have entered an era unlike any before. You may have heard about "big data" and its transformational effect on business operations across multiple industries. Medicine also is entering the era of big data.

It is now possible to have your entire DNA sequence (your genome) decoded, measure tens of thousands of biomolecules in your body, use sensors to continuously follow physiology and activity, and characterize the microbial community that lives in your gut and other parts of your body. These comprehensive measurements have the potential to provide information about your health in exquisite detail, but, as with any big data set, they need to be integrated and interpreted. The hope is that ultimately this information will routinely be used to guide you to manage your health better, not just according to overall outcomes, but based on the experiences of other individuals. In addition to helping decide details about the medical care you get throughout your life, this information has the potential to guide you in many other aspects of your life: the foods you eat, your lifestyle choices, and maybe even the jobs that you choose.

There are many challenges to achieving this goal. Some genomic information we know how to interpret and some we do not; and some of what we think we know will be proven incorrect in the future. In addition, some people worry that despite numerous legal protections against genetic discrimination, genomic information may be used by insurance companies, employers, or others to discriminate against individuals who carry certain genetic markers. Furthermore, there are privacy concerns because it is nearly impossible to ensure that our genetic information will be kept 100% confidential. Some worry that genetic information might be used inappropriately in the social realm; for example, to influence what spouses we choose and whom we encourage our children to marry. Lastly, and perhaps most importantly, there is the challenge of how to pay for genomic sequencing and interpretation. In the United States in particular, healthcare systems primarily provide services to people after they have developed a disease or condition; prevention traditionally has been a lesser focus. Changing this paradigm will require a change in culture.

What personal factors impact our health?

To understand personalized medicine requires an understanding of the factors that contribute to our health (Figure 1). In general, our health is determined by our DNA, our lifestyle, and the things to which we are exposed, namely, our environment. Our DNA is inherited from our parents. Our lifestyle includes behaviors such as how much we exercise, whether we smoke, and what we choose to eat and drink. Most of us know that environmental factors such as the quality of the air we breathe, ultraviolet rays from the sun, and the presence of certain chemicals in household items can affect our health. Important environmental exposures, however, go well beyond these.

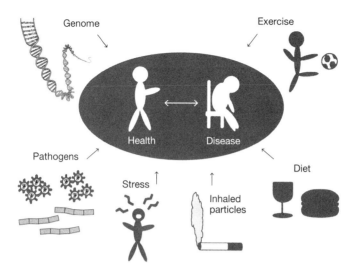

Figure 1 Our health is affected by many factors: our DNA, the various things to which we are exposed (the "exposome"), and exercise.

Pathogens such as viruses and bacteria can make us ill in the short-term, but may also increase our risk of certain chronic diseases in the longer term. Life stresses and the natural processes of aging also affect health. Conditions in the womb can impact development of the fetus and influence health after birth. In many cases, our understanding of the mechanisms by which environmental factors influence health is incomplete. Nonetheless, this information is crucial to understanding how to manage health.

2

GENOME FUNDAMENTALS

When we visit the doctor, we are commonly asked to fill out a questionnaire about our family medical history, including major diseases such as cancer and heart disease experienced by our parents, siblings, and children. This is a rough way of gauging our risk of inheriting the factors leading to these diseases. All this information and more, however, resides encoded in our genome, and the current revolution underway in medicine is to decode and interpret our genomes to predict, prevent, and treat disease on an individualized basis.

What is DNA?

Our DNA is the instruction manual for guiding the process by which we develop from a single cell into a complicated being with many different cell types. We each start as a single cell, created by the fusion of a single sperm and egg. From that single cell, we grow into a mature human being comprised of 30 trillion cells. There are several hundred different cell types (estimates range from 200–300 major categories), each with a different function. For example, our intestinal cells help us absorb nutrients from food, our skin cells protect us from the environment, our retinal cells process light so that we can see, and the neurons in our brain help us to think and communicate. The information in DNA helps guide development and function of all the cells that make a human being.

DNA is a polymer made up of four constituent parts—four nucleotide bases (Figure 2). The nucleotide bases are adenine (A), cytosine (C), guanine (G), and thymine (T). The instructions encoded by DNA are determined by the order in which those bases appear (just as the information in this book is determined by the order in which the words and letters appear). DNA is double-stranded and twisted into a spiral or helical

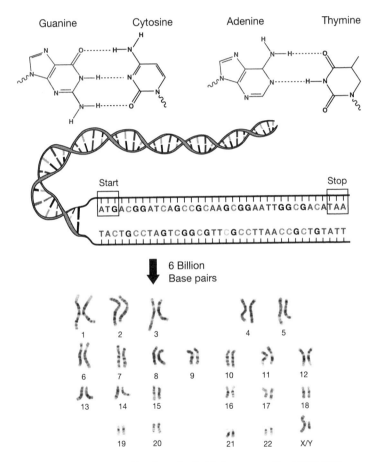

Figure 2. Our DNA is made of four "letters" or (top) that are arranged in a specific order (middle). The bases are paired (middle), and the 6 billion bases are packaged into 46 chromosomes (bottom), 23 from our father and 23 from our mother. Bottom image attribution: By National Cancer Institute [Public domain], via Wikimedia Commons.

shape, hence the common appellation, "the double helix." The sequence of bases on one strand of DNA is complementary to the sequence of bases on the other strand. This arrangement arises because the bases on the two strands are paired: A (adenine) pairs with T (thymine), and C (cytosine) pairs with G (guanine).

What is a genome?

There are 6 billion base pairs of DNA in human cells. The entire 6 billion base pair collection of DNA is the genome. During cell division, the process by which one cell divides to form two new cells, the genomic DNA is copied so that each of the two new cells contains the entire genome. The genome is very long— two meters if stretched end to end—but is tightly packaged to fit into a cell. The genome is contained on 46 chromosomes, 23 of which are inherited from the mother and 23 from the father (Figure 2). Two of the cell types that do not contain the entire genome are the egg and the sperm, which each contain only half the genome or 23 chromosomes.

Our genome contains about 20,000 genes that code for proteins, which are the molecules that carry out many of our biological functions, or in other words, do much of the "work" in our cells (Figure 3). This work consists of processes like digesting our food, storing energy, or copying DNA to generate a new cell. For the most part, each of the various cell types in the body contains the same 46 chromosomes and generally has the same DNA. Cell types differ according to which genes are active, and thus, which proteins are present. For example, our immune cells make proteins called antibodies that bind pathogens and help clear them from our body, retinal cells make proteins that capture light and help us to see, and beta cells in our pancreas make the insulin protein that regulates glucose absorption by cells throughout the body. The regulatory framework that governs which proteins are expressed and at what levels in various cell types is contained in the DNA.

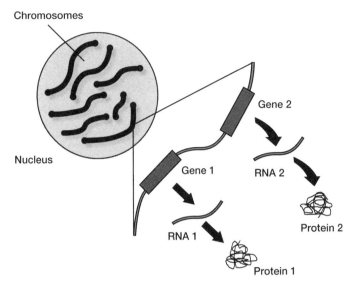

Figure 3. Our DNA encodes the 20,000 genes that make proteins that carry out many of the biochemical activities in our various types of cells. The proteins are not made directly from DNA. Rather our DNA makes a similar copy of itself called RNA, which in turns gets "translated" into forming proteins. Different proteins are expressed in distinct cell types to give cells their unique identity.

Proteins are not made directly from genes but rather through a derived close copy of DNA called "messenger" RNA; the production of this type of RNA is an intermediate step in the production of proteins from genes. In addition to protein-encoding genes, there are probably another 20,000 or more genes that encode other types of RNAs that are not made into proteins. The function of most of these RNAs is not known; many of those with known function appear to have a regulatory role in turning other genes on and off.

Only about 2%–3% of our genome is accounted for by the ~40,000 genes that encode proteins and RNA. It is likely that 10% or more of our genome helps control the expression of genes (i.e., turns them on or off) in the various cell types. The purpose of most of the genome is not known, however, and may not have any direct function.

What does the genome do?

Our genome is solely responsible for many of our traits. These include physical traits such as eye and hair color and ear lobe shape. Our genome is also responsible for affecting behavior and complex diseases such as depression, autism, and schizophrenia. It also affects how we interact with our environment—everything from our sensitivity to air pollution to how we react to alcohol intake and how well we derive nutrition from certain foods. For example, most of the adult world population cannot digest milk because the gene encoding the critical protein for digesting milk (lactase) is normally turned off post weaning. The exception is certain northern European populations and some other groups that do not turn off the lactase gene and thus can typically digest milk throughout their lives. Finally, our genome affects our susceptibility to particular diseases.

The example of lactose intolerance due to lactase insufficiency is an example of a trait due to variation in a single gene (a monogenic trait). Many traits are multigenic, that is, they are determined by the interaction of the products of many genes. Determining the genetic cause of a monogenic trait is relatively easy; determining the genetic basis for a multigenic trait is more complicated. Finally, many traits are influenced not only by an individual's genome but also by his or her environment; that is, traits may have both a genetic component and an environmental component.

How does one person's genome differ from another person's genome?

People differ from one another by several types of DNA sequence changes or "variants" (Figure 4a and 4b). Approximately 3.8 to 4 million variants, or one in every 1,200 bases, are single letter or base changes (termed single nucleotide variants, or "SNVs"). There are also about 500,000 to 850,000 small insertions and deletions of 1–100 bases ("indels"). Finally, there are also thousands of large insertions, deletions,

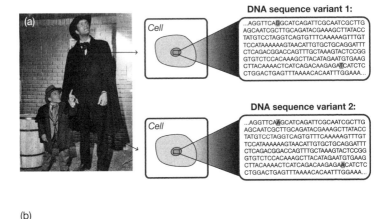

DNA sequence variant 1:

```
...AGGTTCAGGCATCAGATTCGCAATCGCTTG
AGCAATCGCTTGCAGATACGAAAGCTTATACC
TATGTCCTAGGTCAGTGTTTCAAAAAGTTTGT
TCCATAAAAAAGTAACATTGTGCTGCAGGATTT
CTCAGACGGACCAGTTTGCTAAAGTACTCCGG
GTGTCTCCACAAAGCTTACATAGAATGTGAAG
CTTACAAAACTCATCAGACAAGAGATCATCTC
CTGGACTGAGTTTAAAACACAATTTGGAAA...
```

DNA sequence variant 2:

```
...AGGTTCAAGCATCAGATTCGCAATCGCTTG
AGCAATCGCTTGCAGATACGAAAGCTTATACC
TATGTCCTAGGTCAGTGTTTCAAAAAGTTTGT
TCCATAAAAAAGTAACATTGTGCTGCAGGATTT
CTCAGACGGACCAGTTTGCTAAAGTACTCCGG
GTGTCTCCACAAAGCTTACATAGAATGTGAAG
CTTACAAAACTCATCAGACAAGAGAACATCTC
CTGGACTGAGTTTAAAACACAATTTGGAAA...
```

(b)

(1) Single nucleotide variants (SNVs): 4 million/person

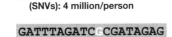

GATTTAGATCGCGATAGAG
GATTTAGATCTCGATAGAG

(2) Short Indels (Insertions/Deletions 1–100 bp): 500K/person

GATTTAGATCGCGATAGAG
GATTTAGA _ _ _ _ _ TAGAG

3) Structural variants:

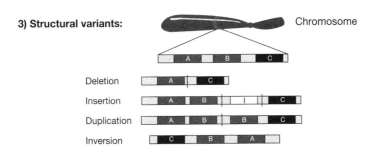

Chromosome

Deletion

Insertion

Duplication

Inversion

Figure 4. (a) Changes in our DNA called "variants" make us different from one another genetically. These differences can lead to the production of different RNAs and proteins, which make us each unique. (Note the DNA sequence is hypothetical and does not correspond to the individuals shown.) (b) There are three general types of genetic variants among people: (1) Single letter or Nucleotide Variants (SNVs), (2) Short Insertions and Deletions (Indels), and (3) Larger and more complex rearrangements called Structural Variants (SVs). These SVs can be Deletions, Insertions (I), Duplications, and Inversions. Left photo: https://upload.wikimedia.org/wikipedia/commons/6/6d/Michael_Dunn_Richard_Kiel_Wild_Wild_West. Right image: National Institute of Human Genome Research.

inversions, and other types of chromosome rearrangements, some as large as several hundred kilobases (1 kilobase = 1,000 bases) in length. These are called structural variants and although fewer in number than single nucleotide variants, they affect many bases and are an important contributor to genetic diversity among individuals.

Many variants appear in the genes that encode proteins. A variety of evidence suggests, however, that most differences that make people differ from one another actually occur in the sequences that regulate gene expression, rather than occurring in the actual protein-coding sequences. These variants contribute to differences among individuals in physical, personality, and disease-susceptibility traits.

With all the variation among genomes, you may be wondering why there are not more dramatic differences among individuals and not more individuals afflicted with genetic diseases. At the DNA level, the average difference between two unrelated individuals is believed to be approximately 0.1%. In a 6 billion base pair genome, this represents a substantial amount of DNA. However, not all differences in DNA sequence are functionally relevant. A variant may occur in a region of the genome that does not contain genes and does not have any critical function. Even if a variant occurs in a protein-coding gene, it may not necessarily affect the expression of that gene or the form of the protein that is produced. Furthermore, because individuals inherit DNA from both parents, for many genes there is a backup system—if the gene from the mother contains a deleterious variant, it may not have a discernible effect if the corresponding gene inherited from the father is fully functional. This backup system may not work in all cases; for example, if the product of the gene with the deleterious variant actually interferes with the activity of the "normal" gene or if proper expression of both maternal and paternal genes is required to achieve adequate gene product for normal function. Finally, if a variant has a profound adverse effect, it may interfere so severely with development that it causes fetal death and that variant, therefore, is not found in living human populations.

How do genomes of men and women differ?

We all have 22 pairs of regular chromosomes, called autosomes, and two special chromosomes called sex chromosomes. Women have two copies of a sex chromosome known as the X chromosome. Men have one X chromosome plus a sex chromosome known as the Y chromosome. The Y chromosome is the smallest of all the chromosomes and contains a limited number of genes, many of which are involved in male sexual development.

Many diseases that only affect men are related to variants in genes on the X chromosome. Because males only have one X chromosome, a variant that causes dysfunction or loss of function of a gene on the X chromosome is more easily manifested in a male than in a female. For example, a common form of red–green color blindness that is over 15 times more common in males than females is due to variants (mutations) in two photopigment genes on the X chromosome. Females have two X chromosomes and thus greater probability of having at least one functional copy of the photopigment genes. Other X-linked diseases that are manifested primarily in males include hemophilia and certain forms of muscular dystrophy.

Interestingly, although females carry two X chromosomes, in any given cell one of those X chromosomes is inactivated, and, as a result, gene expression on that chromosome is greatly reduced. X inactivation is believed to occur to prevent the accumulation of excess, potentially toxic levels of gene products of the X chromosome. As a result of X inactivation, females are genetic mosaics, a term used to describe populations of genetically different cells in the same person. A good representation of female genetic mosaicism related to X inactivation is the mottled fur color of tortoiseshell cats. The mottled pattern reflects the expression of different X-linked fur-color gene products on the two different X chromosomes. Depending on which X chromosome is active, either the orange or black pigment is produced.

How is the genome decoded?

The first whole human genome sequence was completed in draft form in 2001 and in a more finished form in 2003. The project took over a decade to complete, involved ~2,000 researchers, and cost $500 million to $1 billion. The genome sequence is a composite because it was prepared using DNA pooled from several individuals, and thus does not represent a single individual's genome.

The genome sequence was determined primarily using technologies that decoded short fragments, ~1,000 bases in length, and computer-based assembly of those short "reads" into longer continuous sequences based on information from fragment overlaps. The longer continuous sequences of about 150,000 base pairs in length were mapped and assembled into the whole genome sequence based on a previously developed rough map of the genome. The final sequence, referred to as the "reference genome," is approximately 3 billion base pairs in length, and includes each of the 22 autosomes plus the X and Y sex chromosomes. Some gaps in the reference genome sequence still exist; these occur in regions of the genome that are difficult to sequence using current technologies or that are highly variable among people.

Using current technologies, it is now possible to sequence an individual's genome in a few days. The present approach involves sequencing millions of short fragments, typically about 100–150 bases in length (Figure 5a and Figure 5b). The fragment sequences are mapped to the reference genome and variants with respect to the reference are identified (or "called" in geneticists' lingo). The methods have a ~1% error rate. Therefore, to ensure accuracy, each base is typically sequenced an average of 30 times (i.e., the "sequencing depth" is 30-fold

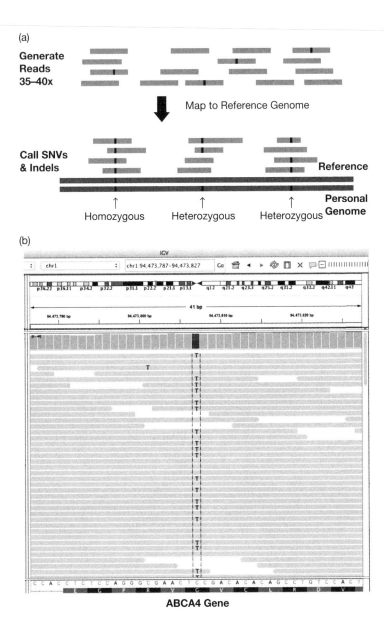

ABCA4 Gene

Figure 5. (a) Deciphering genome sequences. A person's genomic DNA is broken into short fragments and sequences (typically 100–150 nucleotides) are determined from the ends of the fragments. These are mapped to the reference genome and variants (Black Bars) are identified. (b) An example of a variant identified in the *ABCA4* gene, which is involved in retinal function. Mutation in both copies of this gene can lead to retinal disease. Gray bars represent sequences that are identical to the reference sequence shown at the bottom of the figure. Differences in the sequences are shown as letters on the gray bars. This individual has about one half of their reads containing a T instead of a C within the coding sequence and thus is heterozygous, that is, one gene copy has a variant and the other matches the reference sequence.

or 30X). This method works well for calling single base variants and short insertions and deletions. Larger structural variants (large insertions, deletions and inversions), however, are more difficult to determine and specially designed computer algorithms are used to find them (Appendix A). The end result is that a person's genome sequence is defined as variants relative to a reference sequence; and those variants that land in a gene associated with human disease are scrutinized most heavily to assess whether they might be disease-causing.

It is important to note that either one copy or both copies of any given gene (i.e., genes from both parents) can contain a variant with respect to the reference sequence, and it is also possible that different variants can be found in the same or both copies of any given gene. Phasing is a process by which variants are mapped to the same or opposite chromosomes. If two variants occur together on the same chromosome they are "in phase." Phasing is important for predicting the functional consequence of variants in a given gene (Figure 6). For

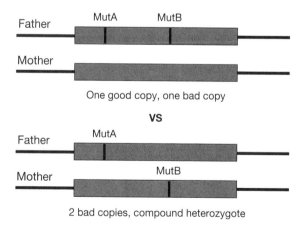

Figure 6. The importance of "phasing variants." When two deleterious variants reside in one gene copy and none on the other, one good copy exists (top). However, if both gene copies carry a deleterious mutation, then both copies are inactivated and this can result in disease. When the mutations are different, this situation is called a "compound heterozygote." Thus, it is important to not only identify variants but know where they lie relative to one another.

example, if sequencing reveals the presence of two deleterious variants affecting a gene that has an important role in taste, and phasing reveals that the two deleterious variants are on the same chromosome (in the same copy of a gene), the copy of the gene on the opposite chromosome will be unaffected and the presence of the normal gene product might mask the effect of the deleterious variants—and the sense of taste will be normal. In contrast, if the two deleterious variants occur in separate copies of the gene (i.e., the genes from both parents are affected), no normal gene product will be produced—and the sense of taste will be compromised.

To date many genomes have been sequenced. A large scale project called the "1000 Genomes Project" has determined the genome sequences of over one thousand people from highly diverse backgrounds and regions around the world. One of the products of this effort is a catalog of many of the common variants found in human populations. Over 50 million variants have been identified. In addition, over one thousand healthy people have had their genomes sequenced, including many individuals who are excited about the potential of personalized medicine and genomics and, therefore, often committed their own resources to have their genome sequence determined. These include celebrities such as Ozzy Osbourne and Glenn Close and many regular people as well. Additionally, many thousands of individuals with diseases such as cancer or diseases of unknown causation have had their genomes sequenced.

There are alternatives to whole genome sequencing that determine DNA sequences of specific, targeted areas of the genome. For example, exome sequencing is aimed at determining the sequences of the protein-coding regions of the genome (Figure 7 and Figure 8). The exome is the portion of the genome that encodes RNA, which is carrying out many of the biological functions—it is the part of the genome that

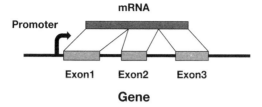

Gene

Figure 7. Genes can be divided into sections, with only some sections represented in the final, mature RNA. The parts of the gene that encode the mature RNA are called exons; the intervening noncoding regions (which are removed or "spliced" out in the final, mature RNA) are called introns. The exome refers to the portion of the genome that is represented in mature RNA. Presently, it is the easiest part of the genome to interpret.

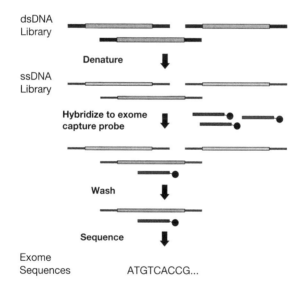

Figure 8. Exome sequencing. Probes can be used to capture only the exome-coding regions of the genome, which can then be sequenced. Because the protein coding exome comprises only ~1%–2% of the total genomic DNA, exome sequencing results in an enormous cost savings compared with sequencing the entire genome. Also, because the scale is smaller, the exome is amenable to deep sequencing, which allows detection of rare variants that might otherwise be overlooked (as can occur in cancer; see Chapter 4).

is easiest to interpret. Other even more targeted approaches determine the sequences of particular subsets of genes that are commonly implicated in certain diseases. These various targeted approaches have historically been less expensive than whole genome sequencing and can provide more accurate sequence information for the regions of interest because "deeper" sequencing is feasible, meaning that the region can be sequenced more times. Exome and targeted approaches may increase the ability to detect heterogeneity (for example, somatic mutations in a subset of tumor cells) in a sample (see Chapter 4). Although the targeted approaches may be more resource- and time-efficient and permit increased depth and accuracy compared with whole genome sequencing, these advantages must be weighed against the lesser amount of sequence information obtained. Also, targeted approaches are limited in their ability to detect structural variants. On the other hand, whole genome sequencing generates a more complete set of information, but its value is necessarily constrained by our current limited knowledge about the non-protein coding regions of the human genome and their role in health and disease. Exome sequencing is presently the most widely used method for decoding a person's DNA, but whole genome sequencing is gaining in popularity.

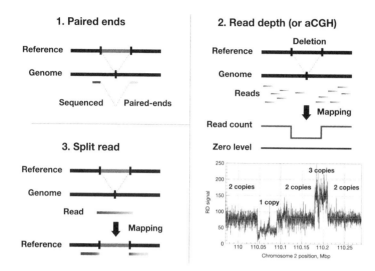

Appendix A. Some of the different computational approaches used for mapping structural variants. (1) By comparing the sequences at the ends of fragments of known length with the reference genomes it can be deduced whether there is a deletion, insertion, or inversion in the sequenced regions. (2) By simply counting the number of independent sequences that map to a genomic interval it can be deduced whether there are normal numbers of copies, too few (i.e., a deletion of one or more copies), or extra copies of a region. An example of a region containing a deletion (dip on the left) and extra copy(ies) (increased signal on the right) is shown beneath the illustration.(3) Sequences that are juxtaposed in a person's genome when normally separated in the reference genome indicate the presence of a deletion in that person's DNA.

3

AN INTRODUCTION TO CANCER GENETICS

What is cancer and how does it arise?

The ability to sequence genomes inexpensively has had enormous impact in medicine and helped usher in the new era of personal genomics. One of the areas of highest impact is cancer, which will strike 40% of men and women at some point in their lifetime.

The ability of each cell in our bodies to grow and divide is tightly controlled by genetic factors; most cells in our bodies either stop dividing all together after we reach adulthood or divide rarely. Cancer results from the loss of these controls, leading to cells that grow and divide uncontrollably. When caught early, cancer often can be managed successfully. If cancer is not detected and treated, however, it may spread to other sites in the body in a process called metastasis. If this occurs, the cancer typically becomes more difficult to treat and can often be fatal. Certain cancers, such as ovarian and pancreatic cancer are often caught late because symptoms are absent or mild until the disease has advanced, and, therefore, these cancers have higher mortality rates.

One of the underlying causes of cancer is genetic mutations that affect cell growth and division, also called cell proliferation.

These mutations fall into two general categories: those that actively drive cell proliferation and those that remove the constraints on cell proliferation. Mutations that stimulate cell proliferation are typically dominant and affect one copy of a gene; the presence of another, normal copy of the gene does not offset or mask the effects of the mutation. These dominant mutations transform a normal gene with cancer-causing potential (called a proto-oncogene) into a cancer-driving gene called an oncogene. We also have a number of genes that encode factors that put the brakes on uncontrolled cell proliferation. These genes are called tumor suppressor genes and generally both copies of the gene must be mutated for a cancer-promoting effect.

In general, development of cancer requires mutations in several different genes—cancers have multigenic causation. This is because there are multiple mechanisms at work in normal cells to ensure that cell growth and division occur at the appropriate times in the proper locations. The body also has defense mechanisms to eliminate certain abnormal cells before they can become cancerous. For most cancers, it is generally believed that mutations accumulate sequentially during a person's lifetime (Figure 9). The person remains healthy until they have enough mutation(s) that wipe out the compensatory mechanisms and thus allow neoplastic, uncontrolled growth. As a tumor continues to grow and the cells rapidly divide, additional mutations may occur that make the tumor more aggressive, for example, or more able to metastasize.

Many different mutation types may contribute to carcinogenesis. DNA mutations may be present at birth or may occur spontaneously during our lifetimes. Mutations that we inherit from our parents are termed "germline" mutations and are generally present in all the cells in our bodies. The term "germline" comes from the origin of these mutations in the germ cells—the sperm or egg. Germline mutations are responsible for the cancers that run in families that lead to high predisposition for cancer.

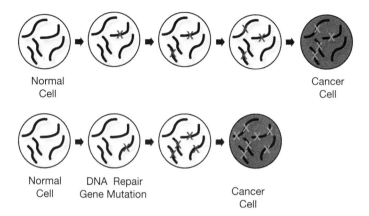

Normal
Cell

Cancer
Cell

Normal   DNA Repair
Cell   Gene Mutation

Cancer
Cell

Figure 9. For many cancers, multiple mutations in proto-oncogenes and tumor suppressors are believed to be responsible for uncontrolled cell proliferation. These are generally thought to accumulate over time. Mutations in DNA repair and chromosome integrity genes such as *MLH* and *BRCA1* are particularly deleterious because they can dramatically increase the rate at which cancer-causing mutations arise.

Mutations that one acquires during one's lifetime are termed "somatic" mutations. Somatic mutations may result when an error is introduced into DNA during cell division or from DNA damage (e.g., damage resulting from exposure of DNA in skin cells to ultraviolet rays from the sun). Somatic mutations are not passed down from parent to child. Germline and somatic mutations can be single base changes, indels, and/or large structural variations.

One type of mutation that is common in many cancers is gene fusion. Gene fusions typically result from chromosome rearrangements that fuse a gene or its regulatory sequence to a proto-oncogene. This results in aberrant, continuous expression or activity of the protein product of the proto-oncogene and uncontrollable proliferation of the cells containing the gene fusion. The first fusion gene to be implicated in cancer was the *BCR-ABL* fusion, which is a hallmark of chronic myeloid leukemia and occurs in > 95% of cases. Several hundred different gene fusions have been implicated in a variety of cancers.

How do the BRCA1 and BRCA2 genes cause cancer?

BRCA1 and *BRCA2* have received a lot of attention because of the link between harmful mutations in these genes and breast and ovarian cancer. *BRCA1* and *BRCA2* encode tumor suppressor proteins. One of the functions of these proteins is to help repair damaged DNA and maintain genomic stability. Women who are born with one defective copy will often lose the function of the remaining copy at some point in time, and this then leads to chromosomal changes and cancer at a very high incidence. The frequency of women with single *BRCA1* and *BRCA2* germline mutations developing breast or ovarian cancer before age 70 is over 80%.

Some women who inherit harmful mutations in *BRCA1* or *BRCA2* opt for surgical removal of their breasts and/or ovaries to reduce their risk of developing cancer. The possible presence of germline *BRCA1* or *BRCA2* mutations may be suspected in individuals with close family members that developed breast and/or ovarian cancer at an early age. Such individuals can get tested to see if they carry a harmful *BRCA1* or *BRCA2* mutation. Genetic testing is helpful but not always conclusive (see detailed discussion in Chapter 8). Sequencing may reveal a mutation in *BRCA1* or *BRCA2*, but it may not be possible to determine whether the mutation is harmful. Furthermore, *BRCA1* and *BRCA2* mutations are not always inherited; sometimes they occur spontaneously, that is, as somatic mutations, and, therefore, may not be detected in the blood or saliva sample used for genetic testing because they are not present in every cell of the body.

Although mutations in *BRCA1* and *BRCA2* are established causes of familial breast cancer, they still only account for 15%–20% of such cases (Figure 10). Mutations in other genes also can be responsible for familial cancer but are present at lower frequency. It is also possible that undiscovered mutations in the regulatory regions of *BRCA1* and *BRCA2* account for additional cases of familial cancer. A recent study of women with a family history of breast cancer, who had normal *BRCA1* and

Other known genes implicated in familial breast cancer

APC	FANCE	PMS2
ATM	FANCF	PRSS1
BLM	FANCG	PTCH1
BMPR1A	FANCI	PTEN
BRCA1	FANCL	RAD51C
BRCA2	LIGA	RET
BRIP1	MEN1	SLX4
CDH1	MET	SMAD4
CDK4	MLH1	SPINK1
CDKN2A	MLH2	STK11
EPCAM	MSH6	TP53
FANCA	MUTYH	VHL
FANCB	NBN	
FANCC	PALB2	
FANCD2	PALLD	

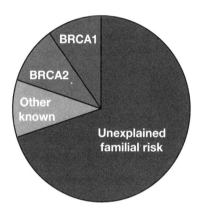

Figure 10. Mutations in *BRCA1* and *BRCA2* each account for approximately 10% of familial breast cancer. Mutations in another 42 genes account for another 10%; the cause of the remaining 70% is not known.

BRCA2 revealed that 10% had mutations in an additional 42 genes. Thus, gene panels that go beyond *BRCA1* and *BRCA2* are likely to be useful in determining an individual's genetic predisposition for breast cancer. Much more work needs to be done, however, to elucidate the genetic basis of breast cancer, particularly for the 70% of familial breast cancer cases for which no candidate genes have been identified.

What are examples of other genes implicated in cancer?

Numerous steps in the normal processes of cell growth and division can become dysregulated due to genetic mutations and contribute to cancerous growth (Figure 11a). There are many genes that, when mutated, can cause loss of control of cell proliferation and contribute to cancer. Examples of some genes commonly implicated in cancer are shown in Table 1.

How does genetic information help us treat cancer?

Genetic information may be useful in tailoring cancer treatment to the specific molecular characteristics of a tumor. Although cancers have traditionally been described by their tissue of origin (e.g., breast cancer, lung cancer, or prostate cancer), in reality cancers of the same tissue may look and behave very differently depending on which mutations are present and which genes are expressed. This is known as the cancer's molecular "signature." For example, breast cancer may be classified into various types based upon which proteins are expressed on the surface of the tumor cells. Breast tumors that express human epidermal growth factor 2 (HER2), estrogen receptor (ER), and progesterone receptor (PR), or are triple negative (do not express HER2, ER, or PR) behave differently and have different prognoses. Tumors that are HER2 positive are treated with medications that bind to HER2 (trastuzumab, lapatinib) and inhibit its activity. ER and PR are hormone receptors, and ER/PR positive tumors are treated with antihormonal therapies. Triple negative tumors have the poorest prognosis and are unlikely to respond to HER2-targeted therapies or antihormonal therapies. Such cancers are usually treated very aggressively with chemotherapy.

As more has been learned about the molecular signature of various cancer subtypes, therapies that are specifically targeted to those signatures have been developed. Conventional chemotherapy acts on all rapidly dividing cells and does not distinguish between cancer cells and normal cells. Chemotherapy may cause substantial side effects, because normal, rapidly dividing cells (e.g., those lining the stomach) are killed along with cancer cells. Radiation therapy is another general approach

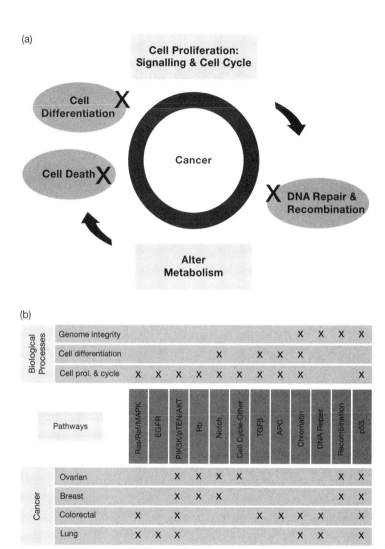

Figure 11. (a) Mutations in diverse biological processes that affect cell growth, cell survival, cell differentiation, and genome integrity can contribute to cancer. (b) Examples of biochemical pathways that are mutated (or altered in expression in the case of Notch) in four common cancers. These pathways participate in the biological processes shown at the top.

Table 1. Examples of Genes Linked to Cancer

Gene	Normal Biological Role	Cancer(s) types in which mutant forms are often implicated
Proto-oncogenes		
PI3K	Cell signalling	Many (e.g., colon, breast, brain, liver, stomach, lung)
BRAF	Cell signalling	Many (e.g., melanoma, ovarian, colorectal, lung, leukemia)
RAS family genes	Cell signalling	Many (e.g., breast, colon, ovarian, lung, pancreatic, leukemia)
HER2	Cell signalling	Breast
Tumor suppressors		
APC	Cell signalling and adhesion; chromosome stability	Colorectal
SWI/SNF complex genes	Higher order DNA structure and gene expression	Many (e.g., ovarian, kidney, liver, melanoma)
TP53	DNA repair; cell death	Many (e.g., ovarian, colorectal, esophageal, head and neck, lung)

to eliminating rapidly dividing cells and has the advantage that it can be directed specifically at the tumor site; however, surrounding normal cells may still be affected. The normal cells that survive chemotherapy and radiation therapy may sometimes acquire harmful mutations that may cause them to become cancerous in the future. The new, targeted therapies attack tumor cells with greater precision and specificity than conventional chemotherapy approaches, and, therefore, can have enhanced antitumor effects and reduced associated side effects.

One notable example of the power of a targeted cancer therapy is imatinib, which is very effective for treating Chronic

Myelogenous Leukemia (CML). Imatinib specifically inhibits the BCR-ABL fusion protein that is constitutively activated in CML. Another targeted therapy, erlotinib, inhibits a receptor involved in cell growth control, the epidermal growth factor receptor (EGFR). EGFR is often mutated in many types of cancer, and erlotinib actually binds tighter and, therefore, is more inhibitory against those mutated forms of EGFR than against normal EGFR. Erlotinib is efficacious in the treatment of tumors that carry those mutated forms of EGFR.

In the above examples, a sample of the tumor (e.g., biopsy) is tested to determine the molecular signature. Testing may be by genetic sequence tests (e.g., for *BCR-ABL*, mutated *EGFR*, or *HER2* gene amplification) or tissue protein stains (e.g., for the presence of ER/PR receptors or HER2 protein overexpression). The results of the testing will guide the choice of treatment—it will be *personalized* for the individual.

Tailored treatment based upon results of molecular characterization of tumor cells is fairly common in oncology. The molecular characterization, however, is usually confined to those factors (aberrant proteins, mutated genes) that are typically associated with a given tumor type. Also, molecular characterization is usually limited to known prognostic factors and a handful of high-frequency "druggable targets," that is, factors common in a given tumor type and for which targeted therapies exist. Rare but potentially "druggable" changes might be present but overlooked. Also, because the molecular characterization of the tumor is limited to already known, well-characterized factors, the information generated has limited value for advancing our overall understanding of cancer biology.

4

GENOMICS AND CANCER TREATMENT

What has been learned from genome sequencing of cancer?

With the ability to sequence whole genomes and exomes, attention has quickly turned to trying to understand the full spectrum of genetic mutations that underlie cancer. Now the genomes or exomes of tens of thousands of cancers have been sequenced and a number of important new insights into cancer biology have been revealed:

1. Every tumor is different and has a different genomic profile.
2. Certain mutations are common in specific cancers. For example, many cancers have mutations in the tumor suppressor gene, *TP53*; many colon and ovarian cancers have mutations in the RAS pathway; and 40%–60% of melanomas have a very specific mutation in the *BRAF* proto-oncogene. Although the genetic basis of certain cancers was known prior to whole genome sequencing, our knowledge of the genes and biological pathways commonly mutated in many different types of cancers has been greatly expanded.
3. Different types of cancers can have mutations in the same proto-oncogene. For example, mutations in the *BRAF* gene commonly occur in melanoma but also appear in

many other cancers such as colon cancer and thyroid cancer.

4. Although there are many different types of cancer, the underlying molecular defects typically affect just a dozen or so processes or "pathways" (Figure 11b). The pathways are often involved in cell growth and proliferation or in repair of DNA damage. For example, breast cancers often have mutations in the pTEN pathway and Rb pathways, which affect cell growth as well as in the *BRCA* genes, which affect DNA repair pathways. Mutations in genes in DNA repair pathways (e.g., *MLH* and *BRCA* genes) can lead to accumulation of mutations in tumor suppressor genes and oncogenes, giving rise to the cancer indirectly.

5. Overall, it is very clear that cancers are best classified not only by tissue of origin (i.e., the conventional classification system), but also by their underlying molecular defects.

How can genome sequencing advance cancer treatment?

Because every tumor is different and has its own genetic makeup, it is ideally suited for personalized medical management based on genomic information. Although not yet a routine part of care, when genomic analysis is performed, the tumor genome is typically sequenced at more than 80-fold coverage, that is, every base is sequenced 80 times on average (Figure 12); many researchers recommend 200 fold or even higher coverage. Normal DNA isolated from blood or saliva of the same patient also is often sequenced. The sequencing might be of the whole genome, the exome, or a large set of genes implicated in cancer (e.g., Foundation Medicine sequences a panel of 315 genes*). The latter approaches (exome sequencing; gene panel

---

* As of October 2014.

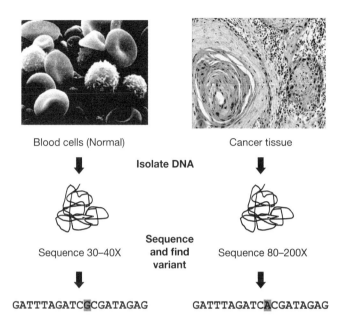

Blood cells (Normal)   Cancer tissue

Isolate DNA

Sequence 30–40X   **Sequence and find variant**   Sequence 80–200X

GATTTAGATC**G**CGATAGAG   GATTTAGATC**A**CGATAGAG

Figure 12. Normal tissue (typically cells from blood) and cancer tissue are sequenced to 30-fold and greater than 80-fold average coverage, respectively. The genetic differences between the samples are compared to find somatic changes (somatic mutations) in the cancer cells. Courtesy of *The Cancer Genome Atlas*, National Cancer Institute and National Human Genome Research Institute.

sequencing) are not as ideal for detecting structural variants, but because they allow deeper sequencing, they may detect variants present in only a subset of the tumor cells. Variants may be present in only a portion of tumor cells because tumors are heterogeneous. Although tumors arise from a single abnormal cell, as the cells in the tumor rapidly divide it is common for new mutations to arise leading to subpopulations of tumor cells with distinct genomic profiles. New mutations accumulate because tumor cells often carry mutations that adversely affect the controls on genome integrity. Deeper sequencing also may provide more sensitive detection of variants in cases where a great deal of normal tissue is present in the tumor sample.

Somatic mutations unique to the tumor cells are revealed by identification of variants (i.e., changes) in the tumor cells compared with the normal DNA. The biggest challenge is to identify which variants are most likely to be "driver mutations," that is, mutations that actively contribute to the growth of the tumor. This task can be especially daunting when whole genome sequencing is performed on tumor cells from advanced cancers because there can be many thousands of somatic mutations compared with the normal DNA from the same person. Indeed, some types of cancer typically have tens of thousands of variants! One typically looks for mutations in known proto-oncogenes or tumor suppressor genes—of highest interest are those mutations that affect targets of FDA-approved drugs or drugs in clinical trials, that is, druggable targets. Examples include EGFR and PDGFR; these are inhibited by erlotinib and sunitinib, respectively. As an example, we sequenced the genome of a metastatic colon cancer patient and found increased copies of the EGFR gene (Figure 13). This patient was treated with an EGFR inhibitor as part of his therapy. Examples of the many targeted drug treatments often used for non-small-cell lung cancer are shown in Table 2.

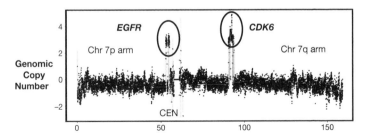

Figure 13. Sequencing of a patient with metastatic colon cancer revealed amplification of the EGFR and CDK6 regions. This patient was treated with an inhibitor of the EGFR signalling pathway. (Hanlee Ji and Michael Snyder)

In addition to finding somatic changes in a person's cancer, usually by sequencing a person's normal DNA, the presence of common germline mutations that increase the risk of cancer is also investigated. Although these germline mutations are usually not "druggable," the information may be useful for determining prognosis and is valuable for family members of the patient in determining their own cancer risk. Indeed, there have now been a number of instances where sequencing the normal and cancer DNA of one family member has revealed a germline mutation that has alerted other family members of their own increased cancer risk.

Genomic analysis might reveal the presence of a potential driver mutation that is not targeted by standard treatment for a given type of cancer. It may be that a drug targeting the effects of that driver mutation is commercially available, but only approved by the U.S. Food and Drug Administration (FDA) for another form of cancer. Or, it may be that a drug targeting the effects of that driver mutation is not commercially available, but is accessible to patients through a clinical trial. The oncologist might use the information from the genomic analysis to select a nonstandard, personalized treatment for the patient. In the scenario described here, the oncologist might use the evidence of the driver mutation to treat the patient with an off-label targeted therapy. "Off-label" refers to the

Table 2. Examples of Drugs Used to Guide Treatment of Non-small-Cell Lung Cancer Based on Genetic Changes. (Reference: National Comprehensive Cancer Network (NCCN) Clinical Practice Guidelines in Oncology.)

Genetic Variations	Targeted Therapy Agents
EGFR mutations	Gefitinib, erlotinib, afatinib
HER2 mutations	Trastuzumab, afatinib
BRAF mutations	Vemurafenib, dabrafenib
MET amplification	Crizotinib
RET rearrangements	Cabozantinib

use of drugs outside of their FDA-approved prescribing information. Physicians have discretion to prescribe off-label, but should have a scientific rationale for doing so. As an example, a drug that targets mutant forms of EGFR and that is approved only for the treatment of lung cancer may be prescribed by an oncologist for colon cancer if testing reveals that the colon cancer cells carry a mutated EGFR and the clinician feels that the patient would be an appropriate candidate for the drug. Physicians' liability risk may be greater when prescribing an off-label use, depending on how strong the scientific evidence is to support that use. Insurance companies and other third party payers (e.g., Medicare) might require documentation from the physician of the scientific rationale before they will agree to pay for off-label use. In an alternative resolution to the scenario described here, the oncologist might use the evidence of the driver mutation to recommend the patient enroll in a clinical trial in which the drug under investigation targets the effects of the driver mutation.

The potential for genomic analysis to help guide personalized treatment for cancer is generating considerable excitement. The hope is that drugs will no longer be used against cancer in situations where they have an extremely low probability of success, and, more importantly, targeting drivers of uncontrolled growth present in a given tumor in a timely manner will be highly effective with reduced side effects. It is important, however, to note that most of these therapies are usually targeted at late-stage cancer patients. Most patients follow the normal course of treatment that has been well established, and many of the drugs used are new and have adverse side effects. As drugs become more specific and their efficacy better known, however, it is expected that these targeted approaches and therapies will be administered to earlier stage patients and become more commonplace.

Another area of active research is the use of genomic and gene expression data to determine when to start anticancer treatment. All anticancer treatments carry a risk of side effects.

For the patient (and the physician), it would be extremely useful to know how aggressive the cancer they are dealing with is so that they can make a well-informed decision about whether to start treatment immediately or delay treatment and just keep a close eye on tumor growth ("active surveillance"). This information would be especially useful in cases where the cancer is quite likely not to be aggressive, whereas treatments may have serious, life-altering side effects. For example, early stage prostate cancer often progresses very slowly over the course of many years, and some treatments may cause incontinence or impotence. Thus distinguishing aggressive versus nonaggressive prostate cancer would be valuable for knowing when and what type of treatment to use for that cancer.

If I have cancer should I get my tumor genome sequenced?

When genome sequencing first became available, many physicians avoided it citing concerns about its accuracy, the complexity of its interpretation, and its limited value in informing treatment decisions. These concerns still persist among some physicians today, but it is hard to imagine that these concerns will persist much longer. Aside from cost, there is nothing to lose by sequencing the tumor genome or at least the genes encoding known druggable targets and prognostic factors—one can, in fact, gain information. The following are possible outcomes of a cancer genome analysis with regard to treatment decisions:

(a) Nothing new will be learned
(b) The analysis will support the patient's current course of treatment
(c) New information will be revealed, suggesting a new course of treatment
(d) Relevant information about an individual patient's likely sensitivity to certain drugs will be revealed
(e) The information will be used for immunotherapy (see below).

We have found that (b) and (c) are frequent outcomes. The genome sequence might confirm that the patient is on the right course of treatment. For example, we sequenced the tumor DNA of a breast cancer patient whose tumor previously tested HER2-positive; the sequence revealed extra copies of the *HER2* gene and confirmed that her current course of HER2-targeted therapy was indeed reasonable. Because initial screening may generate a false positive, the genome analysis provides helpful confirmatory evidence.

An example of scenario (c) might be a breast cancer patient who has completed several courses of chemotherapy, has met her lifetime allowable dose of anthracycline-based therapy (i.e., certain types of chemotherapy are no longer possible for this patient), and tumor genome analysis reveals abnormal activation of a cell signalling pathway that is targeted by a drug currently in clinical trials. If the patient is eligible for one of those clinical trials and the oncologist determines the trial would be appropriate for this patient, the patient could be encouraged to enroll.

Information revealed about an individual patient's drug sensitivity (scenario "d") might include whether that individual would be expected to rapidly or slowly metabolize certain anticancer drugs. Individuals with genetic variants associated with slower metabolism of a given drug will take longer to clear that drug from their body and require lower drug dose (to avoid accumulation of the drug and associated side effects); conversely, individuals with faster metabolism will require higher doses of the drug (to ensure sufficient drug exposure in the body to fight the tumor cells). Finally, if an individual has genetic variants that result in a seriously compromised ability to metabolize a given drug, that drug may be toxic and not suitable for administration to that individual. Presently, drug sensitivity screening before treatment is possible for several chemotherapy drugs (6-mercaptopurine, thioguanine, capecitabine, 5-fluorouracil, and irinotecan) and this list will likely expand dramatically in the future as more genome sequences are determined.

Why do anticancer drugs fail and how might genomic approaches help address this issue?

For advanced cancers, nearly all cancer drugs eventually fail. This can occur at two stages: upon initial treatment and through recurrence of the cancer. Some patients do not respond to a drug initially even though genetic information suggested otherwise. This can be due to additional mutations present in the targets, which confer resistance, or because other cellular pathways are active in the tumor pathway, which counteract the treatment. One example involves EGFR. Patients with a common mutation in the *EGFR* gene (EGFR L858R) are sensitive to erlotinib; however, patients can have a second mutation (EGFR T790M) which confers resistance and thus they do not respond to the therapy. Therefore, understanding the complete picture is very important for treating these patients. In most cases, we do not understand why a set of patients do not initially respond to treatments and efforts are underway to identify the "bypass" mutations and pathways.

One of the major challenges of cancer treatment is that even if treatment makes a tumor stop growing or makes it shrink, as long as the tumor remains, drug-resistance nearly always emerges over time—even in cases in which treatment is highly targeted against a major driver of uncontrolled proliferation, such as mutant BRAF in melanoma. Sometimes resistance appears quickly and sometimes it emerges after several years. The emergence of drug resistance may be understood in the context of tumor heterogeneity. As discussed earlier in this chapter, the cells that compose a tumor are not genetically identical and new mutations arise as cancer cells grow. Additionally, previous chemotherapy efforts may cause new mutations to appear in a tumor that is not completely eliminated. Furthermore, all the cells that compose a tumor may not share the same gene expression pattern (even if they are genetically identical)—for example, differences in where they are located within the tumor may affect which genes are expressed. Certain genetic variants and certain gene expression profiles present in only one or a small number of tumor cells may make those cells less susceptible or resistant to the effects of an anticancer drug. In the presence of the drug, the less susceptible/resistant cells continue to grow and divide. They may even pick up additional mutations that further promote proliferation and/or make them more drug-resistant. Eventually there will be enough of these drug-resistant cells to make tumor growth detectable.

There are a variety of biological mechanisms underlying drug resistance that might be detected by tumor genome sequencing. For targeted drugs, resistance might be related to the presence of a subset of cells that do not express the target because they do not carry the driver mutation and/or carry additional mutations that alter the interaction of the drug with the target. To be effective, many drugs need to be internalized and accumulate within tumor cells. Drug resistance can be related to mutations that disable the mechanism by which the drug is internalized or hyperactivate the mechanism by which the drug is removed from cells. Drug resistance also might be related to genetic variants that activate compensatory mechanisms that offset the detrimental effect of a drug on a tumor cell (e.g., if a drug damages tumor cell DNA, repair mechanisms might be activated).

Combination therapy in which two or more drugs are administered together is common in cancer treatment and helps address tumor heterogeneity and the emergence of drug resistance. This is poorly understood in many cases, however, and efforts are underway to optimize combination therapies. In the future, it will be important to understand the exact genetic composition of each person's tumor, its evolution, and the effectiveness of drug combinations based on this information so that individualized treatments can be prescribed most effectively.

*Can genetics and genomics help detect early cancer
and monitor treatment effectiveness?*

Many cancers, such as ovarian cancer and some kidney cancers, are typically only detected when they are relatively far
advanced, because they may not be associated with noticeable
symptoms in the earlier stages. One remarkable discovery in
the past few years has been the finding that the DNA of solid
tumor cells often can be found in the blood. This presumably
occurs through the death of cancer cells whose contents are
then released into the bloodstream. The discovery of this circulating tumor DNA (ctDNA) is important because it offers the
possibility that a simple blood test might be used for detection
of "silent" early cancers. Indeed, in a recent study designed
to detect regions from 139 genes that frequently carry somatic
mutations in non-small-cell lung carcinoma (NSCLC), ctDNA
was detected in approximately 50% of patients with early
(Stage I) NSCLC and in all of the patients with more advanced
(Stage II–IV) NSCLC. This type of ctDNA assay is theoretically
adaptable to many types of cancer, and it is likely that screening for ctDNA will become a standard component of early
detection in the future for individuals known to be at risk for
certain cancers. This raises the possibility that screening for
cancer lesions using ctDNA in the blood may someday become
routine. Consistent with this idea, it has been recently discovered that analysis of ctDNA in pregnant women for chromosome aberrations in the fetus (see Chapter 9), has revealed the
presence of cancer DNA in the woman! Thus, it may be that
expectant women will be among the first to receive this type
of screening.

Other potential applications of ctDNA analysis are in
patients already diagnosed with cancer—to monitor the effect
of treatment, detect cancer recurrence, and even provide some
insight into prognosis. Some studies have shown that after surgery to remove a tumor, the presence of residual ctDNA is associated with an increased risk of tumor recurrence. Thus ctDNA
screening after surgery might be useful for identifying patients

who should receive chemotherapy to eliminate remaining tumor cells. An increase in ctDNA in a patient whose cancer has been treated successfully could be an early sign of cancer recurrence. Sequence analysis of ctDNA provides a "real-time" look at the mutation profile of the tumor cells and could be helpful in determining prognosis. Finally, ctDNA could conceivably be used for tumor DNA analysis in cases where it is not possible to biopsy or not desirable to subject a patient to the risks of a tumor biopsy procedure. For example, for a patient with NSCLC who is in poor health and not responding to current treatment, and for whom lung biopsy is considered too high risk, analysis of ctDNA could be helpful for determining which somatic mutations are present and informing the decision about alternative treatment. Assessment of ctDNA is often referred to as a "liquid biopsy" because of its potential to perform many of the same functions as a conventional solid tumor biopsy.

Blood is not the only easily accessible material that has potential to be used to detect and monitor DNA released from solid tumor cells. Depending on the cancer, tumor DNA also may be found in stool, urine, and ascites (fluid accumulated in the abdomen in certain tumors). It is possible to detect colorectal cancer DNA in stool samples, and this is now the basis of a commercially available diagnostic test. The utility of urine DNA analysis for detection of early prostate cancer or bladder cancer is an area of active research.

A new approach: What is immunotherapy?

Genomics is beginning to have an impact in an entirely new form of treatment called "immunotherapy" in which a person's immune system is used to attack the cancer. An important factor in developing cancer is the tumor's ability to evade the immune system. Typically, when a cell develops a cancer-causing mutation and becomes precancerous, the immune system recognizes that something has gone wrong with this particular cell leading to an immune response that eliminates the precancerous cells. It is very possible that many of us have developed precancerous cells throughout our lives, but that our immune systems eliminated them before they ever became cancer, and, therefore, we never even knew we had them in the first place.

One trick cancer cells use to evade the immune system is to erect a "shield" around themselves that essentially stops the immune system from launching its attack on them. In some tumors, cancer cells begin massively overproducing signals that essentially tell the immune system, "Do not attack me!" One key signal is the PD-L1 protein. PD-L1 sits on the surface of the cancer cells and tells the immune cells not to attack. It does this by binding to a protein on the surface of the immune cells called PD-1. Recently, it has been recognized that this mechanism for escaping the immune system is much more common in cancers than previously thought.

Immunotherapy is a new major form of treatment that works by blocking the PD-L1/PD-1 shielding system that cancers use to turn off the immune response against them. After turning off those signals, the immune system is free to attack the tumor. Current immunotherapies usually block PD-1 (e.g. Nivolumab and Lambrolizumab), which essentially stops the "shut down" signal from the tumor to the immune system and allows the immune system to attack. Used in conjunction with traditional radiation and chemotherapies as well as targeted therapies, immunotherapy improves outcomes for some cancer patients.

To know whether immunotherapy will be effective, it is important to identify the presence of PD-L1 in the tumor and PD-1 in the immune cells. The presence of these proteins can be determined by analyzing the RNA or searching for the protein (using histological stains). When a lot of RNA or protein for PD-L1 is detected in the tumor, we know that PD-L1 is being created in those cells and that the "do not attack" signal is being sent to the immune system. Such tumors are susceptible to this type of immunotherapy. Anti-PD-1 antibodies therapy is currently being used to treat many different types of cancer (e.g., melanoma, or kidney and lung cancer).

A related immunotherapy strategy that boosts the immune response involves CTLA-4 , a protein that sits on some of our immune cells and suppresses the attack against cancer cells. Antibodies that block CTLA-4 (e.g. ipilimumab) can hyperactivate the immune system and have proven useful for certain types of cancer (melanoma).

How can genomics be used to harness the patient's own immune system to fight cancer?

Although most targeted therapies, such as immunotherapy, use information to identify when a particular therapy will be effective in a particular patient's cancer, another exciting personalized therapeutic approach has demonstrated that strengthening the immune response against the specific tumor can be a highly effective therapeutic approach.

As noted above, each patient's cancer has its own set of unique mutations that happen by chance—most of these mutations are not driving growth of the cancer, but instead are simply passengers traveling along with the driver mutations. Although these mutations are not technically causing the cancer, pieces of the mutated proteins they create (called "neoantigens") can be recognized by the immune system and targeted for attack. By "priming" (i.e., "vaccinating") the patient's immune system against passenger mutations in the tumor (using protein fragments carrying these mutations), the immune response can be magnified, facilitating a stronger attack on the cancer. Recent work has demonstrated the effectiveness of using the patient's own immune system to fight cancer. Promising research in skin cancer and other cancer types is now revealing how neoantigen therapy can further improve therapeutic outcomes for cancer patients.

Each person's cancer has a number of unique passenger mutations. These passenger mutations can be identified through genome sequencing. It is not enough, however, to identify the mutations in the genome; the mutations must also be expressed as proteins in order to create neoantigens. Analysis of gene expression patterns can be used to measure the expression of these mutations, and allow personalized neoantigen therapy to be used to further improve cancer therapy. Once neoantigens that are expressed are identified, the patient can be vaccinated against those particular neoantigens, empowering the immune system to specifically attack the cancer cells. It is expected that this type of approach, which is highly individualized based on a person's mutations, will be extremely effective against many types of cancer and can be used in conjunction with other therapies.

In addition to vaccination against personalized antigens present in cancer patients, another more generic vaccination approach has been described for prostate cancer. Ninety-five percent of prostate cancers express a protein called prostatic acid phosphatase (PAP). By boosting the immune response against PAP, a person's immune response can be specifically tuned to target prostate cancer cells and not normal prostate cells. Thus, immune responses against generic antigens as well as neoantigens can be used to successfully target cancer cells. As with other forms of immunotherapy, these therapies can be used in conjunction with other types of treatments, thereby providing diverse avenues with which to fight cancer.

5

SOLVING MYSTERY DISEASES

What is a mystery genetic disease?

Diseases of genetic origin might cause symptoms at birth or manifest later in life. By age 25, more than 5% of the population has a disorder that has an important genetic component. Thus approximately 16 million people in the United States are affected. In many cases, the underlying genetic variant is not known.

Genetic disorders that cause symptoms early in life often, but not always, are caused by a variant in a single gene. We call this type of disorder "monogenic." Genetic disorders attributable to variants in a single gene follow predictable inheritance patterns as originally described over one hundred years ago by the father of genetics, Gregor Mendel. These disorders are often described as "Mendelian" and can be either recessive (i.e., the copies of the gene from both mother and father each carry a mutation for the disorder to be manifested) or dominant (i.e., only one defective copy of the gene is needed for the disorder to develop) (Figure 14).

Sickle cell disease is an example of a Mendelian, recessive disease. Individuals who inherit the sickle variant of the hemoglobin β-globin gene *HBB* from both their mother and father have the characteristic sickle-shaped red blood cells (our oxygen-storing cells). This shape causes these cells to become abnormally fragile, leading to anemia (low red blood cell count and oxygen deprivation) and a host of other problems. Individuals who inherit a normal β-globin gene from

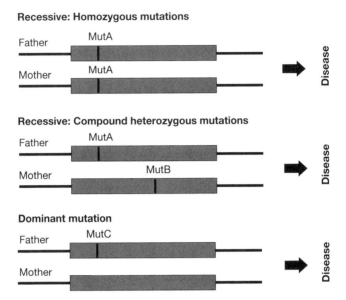

Figure 14. Recessive and dominant mutations. Recessive mutations are those that need to be present in both gene copies for disease to arise. "Homozygous mutations" refers to when each gene copy carries the same mutation; "compound heterozygous mutations" refers to when each gene copy carries a mutation, but the mutations are different. Dominant mutations require only one copy to be mutated to cause disease.

one parent and a sickle variant β-globin gene from the other parent do not have the disease, because the normal copy of the gene is sufficient to supply functional hemoglobin to red blood cells. An example of a dominant Mendelian disorder is osteogenesis imperfecta (brittle bone disease), which is most commonly due to a mutation in a gene for type 1 collagen, an important constituent of bone. Because the mutant type 1 collagen interferes with the function of normal type 1 collagen, an individual inheriting a defective gene from only one parent still exhibits the disease. At one point in time, these diseases were considered "mystery diseases," with the genetic cause unknown; in many cases, decades of research went into discovering the causative genes.

Mendelian disorders may run in families (be familial), or they may arise spontaneously in the germ cells (egg and sperm cells) of the mother or father (or after the fusion of the sperm and egg) and be "de novo" to the child (i.e., not present in the mother or father outside of the germ cells). Thus, when a child with a disorder of suspected genetic origin is born to healthy parents, a spontaneous dominant variant or an inherited recessive variant are the most likely causes. Adding more complexity, disease-causing variants may vary in their "penetrance," that is, in how frequently individuals who carry those variants manifest symptoms of the genetic disease. Highly penetrant variants cause disease in a large proportion (or even all) individuals who carry those variants.

Before the current era of rapid and relatively inexpensive genome or exome sequencing, patients with mystery genetic diseases typically underwent extensive testing and inappropriate treatments before the basis of their disease was understood. It is estimated that it costs $5 million per person to manage these diseases over a person's lifetime. Thus, prompt detection and proper diagnosis of these diseases using new DNA sequencing technologies not only ensures that the afflicted individual receives the appropriate medical care, but also helps realize an enormous savings in healthcare costs. Furthermore, patients and their families are spared unnecessary, stressful, and time-consuming testing. Although not all mystery genetic diseases are solvable using current technologies (see more below), the advent of genomics has dramatically improved success rates.

How many Mendelian diseases are there?

As of July 2014, approximately 7,300 Mendelian diseases have been described. For 3,963 of these, the gene that is mutated and likely responsible for the disease is known (examples are shown in Table 3). It is probable that many more Mendelian diseases will be "solved" as genomic analysis becomes more integrated into clinical practice. For many genes, different genetic variants can have distinct effects on the encoded protein, leading to distinct disease characteristics. For example, different variants may have different effects on the amount of the encoded protein or the activity of the encoded protein. Indeed, the 3,963 unique diseases that have been solved affect only 2,776 genes because different mutations in the same gene can cause different (but often related) disease characteristics.

It is probable that many more Mendelian diseases have yet to be described. There are approximately 20,000 protein-coding genes in the human genome, and variants in many of these genes would be expected to cause human disease. In addition, variants in many of the genes that encode RNAs but do not encode proteins may also lead to disease; however, these have not been studied in detail and the number of known cases is presently limited. It is likely that many more will be found. Of course, some genetic variants may be so deleterious that they result in fetal death and, therefore, will not be observed among the living population.

How are genes responsible for genetic disorders identified?

In general, the ease with which the genetic basis of a disorder may be determined is related to the prevalence of the disease-causing variant(s) in the population at large, to whether there is a family history of the disorder and DNA from other family members is available for analysis, and to the mode of inheritance (e.g., Mendelian, polygenic, and/or high vs. low penetrance).

For example, the cause of certain forms of hearing loss in infants is often relatively easy to determine. This is because hearing loss is relatively common in newborns (one out of every 500 children is born with hearing loss) and is, therefore, well-studied. It is known that two-thirds of hearing loss cases in newborns have a genetic basis. Most newborns with hearing loss (70%) have no other symptoms. When hearing loss is the

Table 3. Examples of Genes Implicated in Mendelian Diseases

Gene(s)	Disease	Possible Management
APP, PSEN1, PSEN2	Early onset Alzheimer's disease (AD)	Medications for symptomatic AD: cholinesterase inhibitors, memantine
BRCA1, BRCA2	Breast cancer; ovarian cancer	For individuals with disease variants but no disease: frequent monitoring for evidence of cancer; in certain cases preventive surgery (mastectomy and/or removal of the ovaries and fallopian tubes) may be considered
CFTR	Cystic fibrosis (CF)	Multiple treatments for symptomatic CF: for example, airway clearance techniques, DNase, ivacaftor for specific CFTR mutations
HFE	Hemochromatosis (excessive iron absorption)	For individuals with symptoms: phlebotomy, chelation, diet modification
WRN	Werner syndrome (premature aging)	Typical treatment for conditions of aging such as hypercholesterolemia, cataracts, diabetes

only genetic disorder present (i.e., there are no other symptoms), and family history suggests the disorder is being inherited in a Mendelian recessive fashion, it is commonly due to variants in one of two genes, *GJB2* or *GJB6*.

In other cases, a much larger but reasonable set of candidate genes can be analyzed. For example, regarding congenital newborn hearing loss that does not have the characteristics associated with *GJB2* or *GJB6* mutations, a set of up to about 100 genes may be examined for variants associated with hearing loss. Through the use of probes that target these genes, the relevant regions can be selected from patients and analyzed. This has proven to be successful in many cases. Understanding the genetic basis of hearing loss may be helpful in determining how to manage the disorder in an affected child. Also, the information may help family members make informed medical and personal decisions.

Targeted approaches that screen a defined set of candidate genes for potentially deleterious variants do not reliably find the causative mutations for many diseases. For example, in inherited hypertrophic cardiomyopathy (HCM), a common cause of sudden cardiac arrest in young athletes, family members may have a strong interest in being tested to determine whether they also carry the mutation underlying this silent, devastating disease. Therefore, identification of the disease-causing mutation is important. However, the disease-causing variant can be detected in only up to 60%–70% of inherited HCM cases using targeted approaches.

Exome or whole genome sequencing often is attempted when more targeted approaches fail. Ideally, the DNA sequences for the entire family including the father, mother, and siblings are determined. Based upon the pattern of inheritance of a disease within a family, the likely genetic basis of the disease (e.g., recessive vs. dominant, familial vs. spontaneous) is deduced. This information is integrated with the results from exome or whole genome sequencing to identify genes and gene variants that might be causing the disease. The signs and symptoms that characterize a disease provide some insight into the likely

underlying dysfunctional biological process(es); ir
available basic research on a disease also helps 1
acterize the underlying biology. This insight into
of the disease can be helpful in narrowing down 1
many genes and their variants identified by the sec
actually causing the disease.

How useful are genomic approaches to solving mystery genetic diseases?

Thus far, for the vast majority of mystery genetic diseases, exome and whole genome sequencing has been successful in identifying the likely disease-causing mutations in about 25%–30% of cases. Furthermore, when the causative mutation is identified, only in a handful of these cases has the information been valuable for optimizing treatment. Often whole genome sequencing or exome sequencing generates a short list of possible disease-causing mutations; narrowing the list further requires additional tests and experiments. There have been, however, a few spectacular successes.

The case of Nicholas Volker offers one early example in which genome sequencing played an important role in the diagnosis and treatment of an unsolved disease. Nicholas was healthy until age two years, when he suffered a cut that would not heal. His condition dramatically worsened, and he frequently developed sepsis (infection of the blood) from many serious wounds, for which over one hundred surgeries were required. He had his colon removed, could not eat or drink, and required total parenteral nutrition (i.e., all his nutrition was administered intravenously). By sequencing his genome, doctors deduced that he had a mutation in the *XIAP* gene, which is associated with immune function. He received a cord blood transplant from an anonymous donor, made a remarkable recovery, and is now doing well at age six years.

Another successful case is that of the non-identical (fraternal) Beery twins (Figure 15). They were born as "floppy"

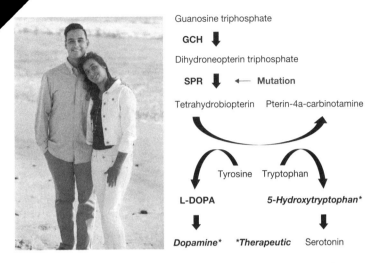

Guanosine triphosphate

GCH ↓

Dihydroneopterin triphosphate

SPR ↓ ◄— **Mutation**

Tetrahydrobiopterin   Pterin-4a-carbinotamine

Tyrosine   Tryptophan

L-DOPA   **5-Hydroxytryptophan***

↓   ↓

Dopamine*   **Therapeutic*   Serotonin

Figure 15. The Beery twins at high school graduation (2015). Using genome sequencing, the twins were found to have a mutation in the *SPR* gene (right). This led to successful treatment with dopamine and a serotonin precursor (5-hydroxytryptophan). Picture courtesy of Retta Beery.

babies, had seizures and delayed motor skills, and were diagnosed with cerebral palsy. Upon receiving a new diagnosis of Segawa's dystonia, the children were treated with dopamine, which improved their health but did not completely alleviate their symptoms; indeed, their health slowly deteriorated. Finally, upon the sequencing of their genomes many years later, a mutation in both copies of the *SPR* gene was discovered in each of the two children. The *SPR* gene is involved in producing both dopamine and serotonin. This finding suggested that in addition to dopamine, serotonin supplementation might be beneficial. Administration of a serotonin precursor to the children along with dopamine significantly improved the health of each child so that they are now symptom-free. Thus, similar to the case of Nicholas Volker, in the case of the Beery twins genome sequencing provided important information that was used for effective treatment and significant health benefits.

Unfortunately, in contrast with these examples
disease-causing mutation does not lead to effective t
the vast majority of cases. Nonetheless, in nearly eve
cases in which a disease-causing mutation is found,
tion is still valuable to patients and their families. Fi
entific explanation typically brings considerable relief to patients
and families and provides some peace of mind. For some individuals and families, the information is helpful for planning
future pregnancies. Some individuals elect to use in vitro fertilization (IVF) and genetic testing to select embryos that lack the
disease-causing mutations. Embryos generated by IVF can be
screened by isolating a few cells from early embryos, amplify-
ing the DNA regions of interest from those cells, and performing
DNA sequencing. Embryos lacking the disease-causing genetic
mutations (or carrying at least one normal copy of the affected
gene in the case of recessive mutations) are chosen for implantation. Thus, although the original genomic information may
not directly benefit the person affected, it can be useful in family
planning. Finally, in some cases the identification of the disease-
causing mutation might be helpful in predicting the possible
course of the individual's disease and their long-term prognosis.

Why can't most Mendelian diseases be solved?

As noted above, in approximately 70%–75% of cases, the
mutation(s) underlying a mystery genetic disease is or are not
found. There are many likely reasons for this impasse. One
important reason is that typically so many variants are present
that it is difficult to narrow in on which one(s) could be causing the disease. By focusing on variants that are likely to be
damaging based on their predicted effect (e.g., inactivation of a
gene product), a smaller number of candidate disease-causing
variants may be obtained, but the number is often still on the
order of five to twenty mutations.

Analyzing larger numbers of affected and unaffected family members can be useful for trimming the list of candidate

disease-causing variants further. In some cases, the clinical features of the genetic disease are suggestive of what underlying molecular mechanisms might be disrupted, and candidate disease-causing variants can be prioritized based on their known/predicted contribution to those mechanisms. Sometimes one of the candidate variants is in a gene that has been extensively studied in mice. If mutation of the mouse counterpart of the human gene results in traits similar to the genetic disease being studied, it suggests that that variant might be disease-causing. However, for patients with a small number of family members available for analysis, and for diseases with more general clinical features, such as developmental delay, it is often difficult to determine which mutation is causing the disease.

A powerful way to identify the likely causative mutation in a rare disease is to compare sequencing information from two or more children whose shared symptoms suggest that they have the same disease. If the children have candidate disease-causing variants in the same gene, there is a high probability that those variants are causative. These "recurrent" rare mutations are extremely helpful because the chances of two children with a very similar rare disease having rare damaging mutations in the same gene and those mutations *not* being disease-causing is very small. Even if the children do not have candidate disease-causing variants in the same gene, if they have variants in genes that have similar roles, then those variants are likely to be disease-causing. For example, many genes implicated in hearing loss contribute to inner ear structure and many genes implicated in cardiomyopathy contribute to the structure of cardiac muscle. If predicted deleterious mutations are found in genes that operate in the inner ear or in cardiomyocytes (which make up the heart muscle), respectively, these might be the likely culprit.

In an interesting example of the power of using recurrent mutations to determine the cause of a childhood disease, we sequenced the DNA of a child with developmental delays and other problems, and her two unaffected parents (Figure 16). The parents had been searching for years to find out what was wrong with their child. After narrowing the list to eight possible mutations, it was unclear which one caused the disease. A second family across the country reported their child with a similar disease and a damaging mutation in one of the eight possible genes (NGLY1). The match was made and the disease solved! This story demonstrates the contribution

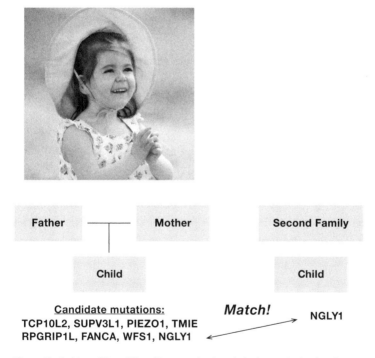

Figure 16. A picture of Grace Wilsey. After screening through the thousands of variants in her DNA, mutations in eight candidate genes were identified. The causative mutation and gene affected was determined once DNA of another child with similar characteristics was sequenced at Duke University and found to have a mutation in the same gene, *NGLY1*. Photo courtesy of Matt and Kristen Wilsey.

of dedicated family members and the importance of sharing information.

In addition to the complexity of sorting through many candidate gene variants, a mystery genetic disease may remain unsolved because no variants were found in the first place. Current sequencing technologies do not cover 100% of the genome. Moreover, different sequencing approaches have different limitations. Exome sequencing analyzes the protein-coding region of the genome, which represents 1%–2% of genomic DNA, hence variants that affect nonprotein coding DNA (e.g., DNA that functions in regulating the expression of a particular gene) are not detected. Also, copy number variations (i.e., whether extra copies of a gene are present) and other structural variations are more difficult to detect with exome sequencing; whole genome sequencing is better for detecting such variants. Exome sequencing, however, targets protein coding regions at much greater coverage than whole genome sequencing (i.e., each base is sequenced an average of more than 80 times for exome sequence vs. an average of 30 times for whole genome) thus, providing greater sensitivity for detecting variants within coding regions.

6

COMPLEX GENETIC DISEASES

What is a complex genetic disease?

Although great strides have been made to identify single gene variants that have a strong causative effect for a particular disease (e.g., *CFTR* mutations for cystic fibrosis and *HEXA* mutations for Tay-Sachs disease), the number of individuals who carry such deleterious genetic variants is overall quite small. More often, single gene changes have only a subtle effect on disease and for most common diseases it is thought that either multiple genetic changes in an individual and/or genetic changes in conjunction with environmental factors cause disease. In contrast to monogenic disorders (also called Mendelian), we refer to those affected by many factors as complex genetic diseases. Examples of complex genetic diseases include type 2 diabetes, autism, schizophrenia, Alzheimer's disease, coronary artery disease and depression.

Our understanding of which genes and environmental factors contribute to a complex disease and how much they contribute to a disease is generally limited. Because the data are incomplete and subject to some interpretation, depending on which institution or company is performing a risk analysis, different genetic risk factors may be weighted more or less (or not at all) in calculating a person's overall risk for a particular complex disease. Furthermore, how individual risk factors interact (i.e., whether their effects are independent of each

rgistic, or antagonistic) is generally poorly under-
ough imperfect, in the absence of more complete
, a simple additive model (based on the assump-
ik factors have entirely independent effects on
disease risk) is often used in determining an individual's combined genetic score for complex genetic diseases (see example
in Figure 17).

For complex genetic diseases, the next frontier in mod-
ern genomic medicine is the identification of all factors (both
genetic and environemental) of small effect that contribute to
an individual's disease and elucidation of how these factors
interact.

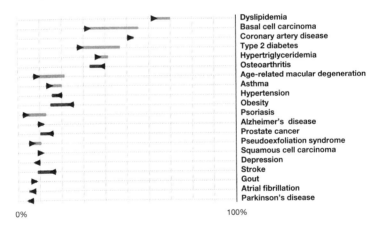

Figure 17. Risk-o-gram of a male (the author) found to be at risk for type 2 diabetes. People
have DNA variants that increase or decrease their chances of developing many common diseases. The triangle is the starting point, that is, estimated risk for acquiring this disease in their
lifetime based on the subject's ethnic group, in the absence of genome sequence information.
By accounting for genetic variant information, increased risk is represented by a bar extending
to the right and decreased risk by a bar extending to the left. Thus, this individual is at higher
risk for hypertriglyceridemia and type 2 diabetes (the patient has both of these conditions), and
decreased risk for obesity (the patient is of normal body mass index). Note that this is a partial
list of results. This figure extends to over one hundred common diseases.

Many neurological diseases are complex
Examples include autism, schizophrenia, c
disorder, and late onset Alzheimer's dise
genes are involved in these disorders like
plex biological processes that underlie human behav...
cognition. There may be many ways to perturb these processes
genetically and environmentally, and, at the same time, there
may be many compensatory mechanisms in place that help
mask the effect of perturbations. Thus, no one individual factor causes the disease by itself, and multiple factors together
contribute to the disease.

Genetic mapping studies, and more recently exome sequencing and genetic tests to search for copy number variation, have
been used to help identify genes involved in various neurological diseases. Many thousands of individuals with autism
or schizophrenia, along with their family members, have had
their DNA analyzed. A number of genes have been identified
that appear to contain an increased number of mutations in certain neurological diseases. Thus far, over 370 candidate genes
have been implicated in autism and over 200 in schizophrenia.
It has also been discovered that gene copy number variations
may contribute to disease and that gene dosage may be an
important contributor to autism. Many of the genes that have
been implicated in autism and schizophrenia are involved in
synaptic functions; synapses are the junctions through which
nerve cells communicate with one another and thus are important for proper functioning of the brain. Interestingly, many of
the genetic loci implicated in autism and schizophrenia over-
lap, suggesting that these two diseases have some common
features. Much effort is underway to identify additional biological pathways associated with each of these diseases.

Recently, research from the author's laboratory utilized a
novel approach to identify a molecular network underlying
autism spectrum disorders. Essentially, existing information

out which proteins like to work together through their physical interactions was used to create an "interactome map." Genes previously implicated in autism were located onto this interactome and many of them mapped onto a particular molecular network or module. The expression pattern of the genes in this autism module was predicted by consulting a brain atlas that details which genes are expressed where in the brain. In addition to cortical neurons, the autism module genes are largely expressed in the corpus callosum and oligodendrocyte cells, a region of the brain and type of cell not usually considered to be linked to autism. Thus, this novel integrated omics approach suggested a molecular network (rather than just a gene or genes) underlying a complex disease and provided insight into how it functions. It also offered a new approach that can be used to help decipher other complex genetic disorders.

Although extensive effort has been devoted to understanding the genetic basis of neurological diseases, as with other complex diseases, to date there is not a strong predictive genetic test to help identify individuals at high risk. If we had this information, we might be able to screen subjects at high risk and detect disease earlier—and possibly improve outcomes with early intervention. In fact, despite the large number of genes that have been implicated in complex neurological diseases, less than 10% of the genetic or other factors underlying any of these diseases have been identified. It is likely that additional genes remain to be discovered. Furthermore, combinatorial effects of multiple genes (i.e., gene–gene interactions) and gene–environment interactions contribute to many complex diseases, and integrating these factors into the determination of disease risk is in its infancy.

How does complex genetics affect metabolic diseases?

Complex metabolic diseases such as diabetes and obesity affect a staggering percentage of the population and a great deal of scientific effort is being devoted to unraveling the complex interplay of genes and environment that cause such diseases. The onset of type 1 diabetes is often in childhood or young adulthood and involves the loss of beta islet cells that produce insulin; genes that have been implicated are involved in immune function (e.g., HLA) and beta cell function. Patients with type 1 diabetes often produce antibodies that target and attack proteins on the beta islet cells, thus contributing to beta cell loss. Type 2 diabetes is caused by progressively increasing dysfunction in the body's response to the glucose-regulating factor, insulin. Although type 2 diabetes is strongly linked to obesity, there are significant genetic factors at play. Not all obese individuals will develop type 2 diabetes, while many lean individuals will develop the disease, and we currently have a poor understanding of what makes these individuals unique. Obesity itself is highly complex, and although diet and exercise are clearly contributing factors, recent research has implicated genetic factors and even the composition of the microbial population in the gut. A third, less common form of diabetes exists and is referred to as MODY for Maturity Onset of Diabetes in the Young. It is often linked to mutations in one of a small number of genes (~10) that can cause disease by reducing the ability of the pancreas to produce insulin. MODY is monogenic and is not a complex genetic disease.

Large genetic mapping studies spanning thousands of people have been performed for analyzing type 1 and type 2 diabetes, obesity, and other disorders such as elevated cholesterol. These have identified over one hundred genes associated with these various diseases. As with complex neurological disorders, in each case, although many genes have been identified, the genetic contribution of these genes in total is estimated to be 10%–20%, depending upon the disease. Thus, additional genes remain to be discovered. Importantly for each of the diseases, environmental factors and pathogen infections contribute to the disease, as described later in the gene–environment and infectious disease chapters. Thus, it is likely that gene–gene and gene–environment interactions contribute to each of these diseases. Overall, because of the myriad of factors that contribute to complex diseases, each person will have his/her own risk for these diseases, based on his/her personal genetic and environmental contributions.

Can some diseases be both monogenic and complex?

It should be noted that for many, and probably most diseases, there is a spectrum of genetic causes from monogenic/ Mendelian to complex. That is, some highly penetrant variants can be strongly associated with a disease, along with other variants of low effect. Examples include cancer, Alzheimer's disease, and heart diseases in which both monogenic and complex forms occur in different families. These forms can have different disease effects. Take as an example Alzheimer's disease. There are early onset cases in which particular variants in the *APP* or presenilin-1 or 2 genes almost always cause familial forms of the disease. That is, if a family member has the disease and the child gets the same variants, there is a good chance the child will develop Alzheimer's disease at a relatively young age.

For late onset Alzheimer's, the situation is more complex. One major gene associated with late onset Alzheimer's is the apolipoprotein E gene (*APOE*). Twenty-five to thirty percent of the population has a variant called ApoE4 that is commonly associated with the disease. (Males without this mutation have a 17% chance of acquiring the disease; those with one copy have a 25% chance; and those with both copies mutated have a 60% chance.) Although the ApoE4 allele is present in about 40% of people who have Alzheimer's, it is not the only factor responsible for the disease, because many people who have two copies of this allele do not develop the disease, and similarly, many others who get the disease do not have the ApoE4 allele. Thus, other genetic and environmental factors contribute to late onset Alzheimer's, and many other genetic loci of lower effect have been identified. Interestingly, in regards to ApoE, there are also alleles that are protective (ApoE2) and reduce the chances of getting Alzheimer's. Thus, mutations in various genes can cause familial disease, and major and minor genetic risk factors exist.

Two other examples of complex diseases or syndromes with a major genetic component—as well as other minor alleles—are age-related macular degeneration and autism. Age-related macular degeneration results in blindness, and one major locus (*HTRAX*) accounts for approximately one-half of genetic cases. For autism, there are mutations that cause strong autism-like effects in children (e.g., Rett syndrome and Angelman syndrome in children, which have mutations in the *MECP2* and *UBE3A* genes, respectively). There are also many other mutations that likely do not cause the disorder by themselves but likely do so in conjunction with other genetic mutations or environmental factors. Thus, overall, for most diseases there is likely to be a spectrum of risk genes, stretching from those that give very strong effects to those that result in much weaker ones.

7

PHARMACOGENOMICS

How can your genome directly help guide drug treatments for treating disease?

As noted earlier, genomics can have a direct impact on drug treatments for cancer and mystery diseases. In a patient with cancer, it may be possible to use that patient's DNA sequence to guide treatment to chemotherapy or to target pathways that are genetically (or epigenetically, as we will discuss later) affected in the patient. The same is true in rare cases for undiagnosed diseases—treatments can be matched to genomic information as in the case of the Beery twins described previously.

What are other ways your DNA can guide drug treatments?

Our DNA also affects drug metabolism, side effects, and drug–drug interactions. This information can be extremely valuable because a wrong drug dose or dangerous side effect can have life-threatening consequences. This area has been studied extensively, and to date there are several hundred identified genes that affect drug response. Many of these genes encode proteins such as the cytochrome P450s, which normally modify natural compounds to make hormones or help remove toxins in our bodies. We have 80 *CYP* genes encoding cytochrome P450s in our genome. These proteins affect the metabolism of drugs we take, either by directly modifying the drug, or by modifying prodrugs (precursors to drugs) that are administered to patients, and thus affect the amount of drug that is available to act in the body. Other genes that affect drug response encode transporters that either deliver drugs into our cells in order for them to work or evict them from the cell so they cannot reach their intended target.

One of the best studied cases of genes affecting drug dose is that of warfarin (also known as Coumadin). Warfarin is an anticlotting agent that is administered to patients with existing or heightened risk of forming clots in their vasculature or to prevent clot formation in their hearts due to cardiac arrhythmias or mechanical valve replacements. The metabolism of warfarin is affected by two genes, *VKORC1* and *CYP2C9*. The -1639G>A SNV in the promoter of the former leads to less protein produced, and so less warfarin is required to thin patients' blood adequately. For *CYP2C9*, the variant forms *CYP2C9*3* (I359L) and *CYP2C9*2* (R144C), metabolize the drug more slowly; patients with these variants are also typically administered lower doses. Thus, genetic screening should theoretically be useful for administering the proper drug amount and help reduce morbidity and mortality from excessive blood thinning and consequent uncontrollable bleeding.

Another example is tamoxifen, which is used for treatment of endocrine responsive breast cancer. Tamoxifen is given to patients postsurgery and dramatically reduces the rate of cancer recurrence. This drug is metabolized by cytochrome P450 2D6, the product of the *CYP2D6* gene. Based on their DNA, there are patients with little CYP2D6 activity who are poor metabolizers and others with high activity who are extensive metabolizers. An FDA-approved genetic test exists for finding the variants of the *CYP2D6* gene to help guide tamoxifen administration, but the lack of study data demonstrating its role in improving patient outcomes has, to date, led insurance companies to refuse to cover the test.

Beyond having ramifications for drug efficacy, genetics also may play a role in the side effects of drugs. Many patients with atherosclerosis take statin drugs. A known but relatively rare side effect of this class of agents is a sensation of muscle burning (myalgia) due to muscle breakdown (rhabdomyolysis). Multiple mutations have been found that modulate one's susceptibility to this particular adverse effect. One mutation appears to be specific for a certain, widely used statin drug, so a patient's genetic information could, in theory, be used to choose which drug he/she should take. Interestingly, the medical society's approach to these findings has been to recommend against placing any new patients on the highest doses of this statin, an across-the-board solution that otherwise could be individualized with genetic information.

It is important to note that nearly all drugs have side effects. Thus, altering doses often goes hand in hand with altering side effects. For these reasons, it is useful to incrementally increase the drug dose to the optimal level for therapeutic response, but not so high as to cause unwanted side effects.

Finally, though it seems obvious that drug choice and dosing can and, therefore, should be tailored to a person's genomic sequence, there are presently few trials that substantiate the benefits of genetic testing for drug treatments. Furthermore, even in the most obvious cases for which the benefits should be evident, such as with warfarin, the trials sometimes do not support the value of genetic testing. The reason for this present situation is not clear, but how the trials were performed may have played a role.

Are there sex differences in drug effects?

A major genetic difference present in all populations is the one responsible for dividing individuals into the two sexes. Men and women express different drug metabolism enzymes at different levels; that is, many cytochrome P450s are differentially expressed in the liver or kidneys of men and women. In addition, men and women have different body masses. Therefore, it is not surprising that men and women often react differently to drugs. Antihypertensive, antipsychotic, and antidepressant

Table 4. Examples of Different Responses of Men and Women to Drugs

Drug	Sex Differences
Statins	Females have greater tendency for myopathy
Verapamil	Females have greater blood pressure decrease
Amlodipine	Females have greater blood pressure decrease
Aspirin	Females have greater risk reduction for stroke but less for myocardial infarction

drugs all have been shown to have different effects in men and women (Table 4). Surprisingly, however, sex differences have only recently been considered in research studies investigating clinical efficacy.

8

GENOMICS FOR THE HEALTHY PERSON

Would knowing your genome affect the disease you might get, the job you choose, the sports you play, and the food you eat? Here we explore the role of knowing your genome sequence if you are healthy.

How can getting your genome sequenced improve your health?

Although genomics can be useful for people with diseases, can it be useful for the healthy person? Many people believe that the answer is no. They feel it is too hard to predict disease risk from this information and raise concerns about the accuracy of both the technology and interpretation. They also worry that people will receive the information and worry excessively about their potential disease risk.

Others, however, have a different opinion. They feel that useful information can be extracted from a person's genome, and that this can be used to help guide a person's health care. A compelling argument is that family history is widely used in medical care. Based on family history, people are commonly placed on the alert for certain diseases, and diet and physical activity programs are often recommended. Shouldn't a person's genome sequence be better than family history?

The answer is certainly, "Yes," even if we are not perfect at interpreting a person's genome. We all have a multitude of

NA, many of which may cause disease—that
has the perfect genome. In fact, we all have at
that cause a gene to be inactivated, although
these cases a good copy still exists. Methods
ished to analyze a person's genome to find
re predictive of disease. A version we use is
summarized in Figure 18. By analyzing a person's genome in
detail we find several types of mutations:

(a) Those that are in a gene that previously has been shown
to cause disease at very high occurrence. Furthermore,
the actual mutation is very likely to interfere with the
normal function of the gene and cause a disease. The
genome is searched for all variants that cause disease at
high frequency. Particular attention is devoted to variants in genes that may cause disease based on a person's
family history,

(b) Those that lie in a gene known to cause a disease, but the
mutations are new or rare. Computer algorithms suggest
the mutation may be damaging, but it is unclear whether
these mutations are disease-causing. These mutations are
called "variants of unknown significance" (or VUSes).

(c) "Carrier mutations" in which the mutation itself is
unlikely to cause disease but could be passed on to cause
disease in a person's children and grandchildren (e.g.,
recessive trait or lowly penetrant dominant one).

(d) Pharmacogenetic variants, which predict the individual's
response to the level of a drug or potential adverse side
effect for a particular drug (discussed earlier in Chapter 7).

(e) Finally, it is possible to estimate the risk of complex diseases by aggregating the individual contributions of the
many genetic loci that have been associated to the disease.
This analysis forms the Risk-O-Gram described earlier.

Examples of mutations that lead to disease with very high frequency are known disease-causing mutations in the *BRCA1*,

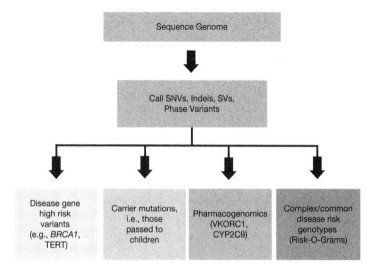

Figure 18. Strategy for analyzing a healthy person's genome. Variants identified by genome sequencing are analyzed for those that (a) reside in genes in which mutation in just one gene copy can cause disease; (b) are passed on to children and when combined with a mutation in the second gene copy may cause disease; (c) may affect response to certain drugs; and (d) have small effect but in aggregate with other changes can be associated with complex disease. For scenario a and b, sometimes the variants are "known" to be associated with disease and sometimes they occur in the same gene but are new and their significance in causing disease is not clear. These variants of unknown significance (VUS) can be difficult to interpret.

BRCA2, and *SHDB* genes and the early Alzheimer's genes (e.g., *APP*, presenilins). For *BRCA1* and *BRCA2*, there are many known disease-causing mutations in these genes; women harboring these variants have a greater than 80% chance of developing breast or ovarian cancer. For mutations in *SHDB*, there is a high chance of developing paraganglioma. Although one might think that these mutations should be evident from a person's family history, this is not always the case. In fact, in a recent study that we performed, a woman who did not have a family history of breast cancer was identified with an inactivating mutation in *BRCA1*. Although the reason for this is not clear; it is possible that she inherited the mutation from her father who is less susceptible to the disease or the mutation is "de novo." This person learned about the mutation from her genome sequence and had surgery based on that information. She would not have known about this mutation had she not had her genome sequenced.

In addition to the *BRCA, SHDB,* and Alzheimer's disease genes, many other mutations are highly predictive of diseases. The American College of Medical Genetics has identified 56 genes for which it is recommended that known disease-causing mutations from a sequencing study be reported back to the subject's physician, who, in turn, can consult with the patients based on their wishes. This list can be easily expanded to include many more genes with known disease-causing mutations and will continue to grow as more disease-causing mutations are found.

"Variants of unknown significance" (VUS) is the second category of mutation. Our work indicates that people typically have one to three of these that might directly contribute to disease, and the number is even higher if carrier status is considered. It is difficult to know whether these mutations will cause disease. Their identification, however, can lead to follow-up tests that can determine whether the person might be likely to acquire this disease. For example, in the author's genome, a variant in a *TERT* gene suggested that he may be susceptible to aplastic anemia, a loss of blood cells. The activity of the gene (which adds sequences to the ends of chromosomes) and levels of blood cells were analyzed in subsequent tests. The subject was found to have slightly shorter sequences on the ends of his chromosomes; however, his levels of blood cells have been normal and thus, do not currently manifest disease symptoms. For diseases that are associated with increased age, patients who have VUS can be on guard for acquisition of these diseases.

Finally, it is possible to infer risk of complex diseases from variants of low effect. In most cases, these may increase a person's risk from a very low value, such as 0.1% to a higher, but still overall low value, such as 1%. Although this is a 10-fold increase, the person is still at very low risk for developing the disease. Presently, the accuracy of these tests is difficult to ascertain. However, they often match family history. Moreover, in the author's case a prediction was made that proved to be accurate. My genome sequence predicted that I am at risk for type 2 diabetes, a condition not known to run in my family. I acquired the disease after a respiratory infection, and because of awareness of my genetic predisposition, detected it early and successfully managed it, at least initially (Figure 19). Like VUSes, complex disease risk can be a useful prognostic indicator of diseases for which to be on the alert, much like family history.

Genome sequencing for the healthy person already has been shown to have value in a number of cases: for predicting disease risk, catching disease early, and avoiding drugs that may lead to adverse side effects. As such, it can help facilitate a shift in the way medicine is currently practiced, from treatment after manifestation of the disease to taking a more

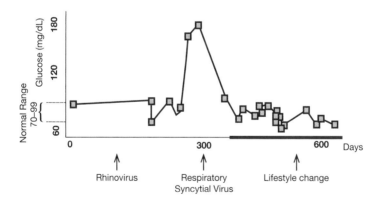

Figure 19. My genome was sequenced and I was found to be at risk for type 2 diabetes based on this Risk-O-Gram. My levels of glucose and glycosylated hemoglobin (HbA1c; an indicator of steady state glucose) were followed—elevated levels of glucose and HbA1c (not shown) were evident after a respiratory virus (RSV) infection. High glucose was controlled through diet and exercise initially, although elevated glucose and HbA1c levels returned two years later (not shown).

preventative approach. Moreover, as our interpretation of the genome improves, the value of genome sequencing will continue to increase further.

Can genome sequencing affect the drugs someone takes?

Even if a person is generally healthy, he or she may take medications on occasion or even regularly. As an example, a young woman had her genome sequenced and was found to have a variant that put her at risk for thrombosis. Based on this information, it is recommended that she should avoid certain oral contraceptives. In a second example, an individual with heart palpitations was found to have a variant responsible for an aberrancy in the electrical properties of the heart called long QT syndrome. It is recommended that this individual avoid certain drugs in the future, including some commonly prescribed antibiotics. Because we already know so much about the genetic effects on drug responses, much more can be learned about drug dosages and what drugs a person should avoid.

Analysis of the genomes of a number of people reveals that a person's genome sequence typically has valuable information regarding about three to six drugs (Table 5). As an example, analysis of the author's genome revealed useful information about several drugs related to type 2 diabetes: metformin, a widely used diabetes drug, and troglitazone, which is no longer prescribed in the United States. Both these medications were predicted to have a greater effect than would be expected for the average patient. The genome sequence also revealed useful information about statins that should not used so as to avoid side effects. Finally, it revealed the previously mentioned *VKORC1*-1639G>A mutation, which increases the blood thinning effects of warfarin. Although it is possible that some or none of these drugs will ever be used, the information may be valuable in the future.

Table 5. Examples of Variants Predicting Drug Response in a Personal Genome. *The variants listed in the Table were found in the genome of the author who has elevated glucose and are potentially relevant for his drug dosing should they ever be needed*

Gene	SNP	Patient Genotype	Treatment	Drug(s) Affected
CDKN2A/2B	rs1,08,11,661	C/T	type 2 diabetes	troglitazone (increased beta-cell function)
CYP2C19	rs1,22,48,560	C/T	atherosclerosis	clopidogrel (increased activation)
LPIN1	rs1,01,92,566	G/G	type 2 diabetes	rosiglitazone (increased effect)
SLC22A1	rs6,22,342	A/A	type 2 diabetes	metformin (increased effect)
VKORC1	rs99,23,231 (or -1639G>A)	C/T	atherosclerosis	warfarin (lower dose required)

Can genetic testing be used to predict sports performance and injuries?

Genetic loci have been associated with sports performance and injuries. Physical endurance, strength, and other building blocks of athleticism are complex traits to which genetic loci have been associated. A gene that directly affects contractile muscle, alpha-actinin-3 (*ACTN3*), may affect strength, and variants in this gene have been associated with sprinters.

Genes involved in energy utilization and hypoxia (i.e., the lack of oxygen) have been shown to be associated with physical endurance. Many of them affect metabolic and mitochondrial function. These findings are not unexpected, given that mitochondria are involved in energy utilization and production. Examples of "endurance genes" include the *PPAR-delta* gene, which encodes a key regulator involved in energy utilization in which a variant (294T/C) has been associated with endurance; *GYS1*, which gives rise to a skeletal muscle glycogen synthase; and the beta2-adrenergic receptor gene. Similarly, variants in genes associated with hypoxia, a response to low oxygen, have been associated with endurance as well. Presumably, these variants affect efficient energy utilization and oxygen consumption in muscles subjected to long periods of use.

Genetic traits also have been associated with sports injuries. Tendon disorders have been associated with blood groups and collagen genes (a structural component of tendons), and collagen variants have been associated with propensities for ligament tears. Individuals at risk for cardiomyopathy can potentially die from physical exertion—indeed, in a high profile case, a rising star for the Boston Celtics, Reggie Lewis, died during practice and was diagnosed with hypertrophic cardiomyopathy, a disease known to have a genetic basis. Recently, concerns have been raised about head blows in football and boxing as causes for Alzheimer's and Parkinson's diseases. The advantages of screening for those at risk for such conditions seems obvious; those at risk for injury or death might choose not to pursue such recreational or career activities, or they might modify their training to minimize their risk. Many individuals who desire to pursue these activities, however, may opt out of testing for fear they could not then pursue them.

Will sequencing my genome affect my children and my relatives?

We all have genetic mutations, and your genome sequence will reveal them. It not only predicts your risk for disease, but indirectly affects your relatives as well. Your children will inherit one half of your damaging variants from you and another half from your spouse (which half is not known, but it can be tested). Likewise you share half of your DNA with that of your biological brothers and sisters and half with your mother and father. Thus, what you learn about you provides some information about disease risk for other related family members. In this sense, it is very much like family history, in which genetic diseases run in families due to shared DNA.

It is important to note, however, that a child's risk for a particular genetic disease is not identical to that of his or her parent because only part of the genetic information is shared between parents and their children. The genetic predisposition loci randomly pass to children so the child may not get the mutation. Thus, just because a parent has a genetic disease does not mean the child will get the disease. A good example is the *BRCA1* mutation. If a mother has this mutation, there is a 50% chance her children will inherit it. Thus, the chance is greater than the population as a whole (normally present in about 1 in every 400 people) but not a certainty.

9

PRENATAL TESTING

*How are genome sequencing technologies changing
prenatal testing?*

One area where genomics is revolutionizing current practice is
prenatal DNA testing. Prenatal DNA testing is a common practice
used to monitor whether fetuses may have abnormal chromosome numbers (aneuploidy), which occurs about once in every
160 live births. It can also be used to identify mutations in genes.
Prenatal DNA testing is commonly used when the chances of
genetic abnormalities are high due to family history or advanced
maternal age (e.g., greater than 35 years). Down, Edwards, and
Patau syndromes are examples of having extra copies of chromosome 21, 18, and 13, respectively, and are the most common auto-
somal aneuploidies. Turner syndrome results from the lack of
the second copy of X chromosome in women. The consequences
of aneuploidies vary widely among the various syndromes and
with the affected person but can include intellectual disability,
congenital heart defects, and infertility, among others.

Until relatively recently, nearly all prenatal testing was performed by amniocentesis or chorionic villus sampling (CVS).
During amniocentesis, a needle is used to collect fluid harboring fetal cells from the amniotic fluid surrounding the fetus
whereas CVS samples the placenta. The cells are then analyzed
for alterations in chromosome number or structure. Both of
these tests are invasive, however, and carry a risk, albeit low,
of miscarriage.

A more recent test involving genomics, noninvasive prenatal testing (NIPT), is now revolutionizing genetic testing of the fetus. It has been known for some time that fetal cells circulate in the mother's blood; later it was discovered that fetal DNA also can be found in the mother's blood. The amount is very low in early pregnancy—approximately 3%–4% of the cell-free DNA in the blood is derived from the fetus at five weeks, and the fraction rises to 10%–15% by the eighth month of pregnancy. With the advent of high-throughput sequencing, it is now possible to analyze the fetal DNA from the maternal blood—the paternal sequence in the fetal DNA has variants that are distinct from the maternal DNA. Originally, this approach was established to identify copy number variations (CNVs). Detection of CNVs is particularly important because births with aneuploidies are quite common, and duplication or deletion of genes can have developmental and/or health consequences. The aforementioned aneuploidies are relatively easy to detect using high throughput sequencing (Figure 20). In fact, a prenatal diagnostic test has recently been approved by the Food and Drug Administration (FDA) for detecting

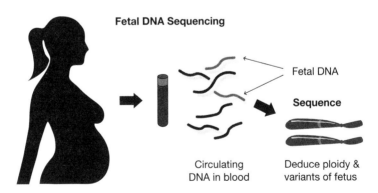

Figure 20. General strategy for genome analysis of circulating fetal DNA. Fetal DNA circulates in pregnant women throughout pregnancy. Total DNA is extracted from the blood fluid (i.e., plasma), sequenced, and the fetal DNA sequence deduced. Presently, this test is approved for detecting aneuploidies in the fetus, but the technology can be used to sequence entire genomes before birth.

trisomies of chromosomes 21, 18, and 13 in a mother's blood. The method is highly accurate and has a sensitivity of greater than 99% with few false positives. Smaller chromosome abnormalities also can be detected with this technology, although such tests are not yet approved by the FDA.

The advantage of noninvasive fetal DNA testing is that it is very simple and can be performed using blood drawn during a routine prenatal checkup. It is only a matter of time before this test will be available for all pregnant women and may become routine. It is likely that some women may decline the noninvasive fetal DNA test, but it is easy to envision that a large number of parents will opt for the test to gain insight into their child's likelihood of having a severe genetic disease.

Can genome sequencing be used to identify other mutations, beyond chromosomal abnormalities, that might cause disease?

With the higher sensitivity and lower cost of high-throughput sequencing, it is now possible to determine the whole genome sequence of a fetus before it is born. Although it is likely that genome sequencing of fetal DNA obtained from the mother's blood would be less accurate than that of fetal DNA from the fetus itself, prenatal genome sequencing nonetheless provides an opportunity to noninvasively predict disease risk before a child is born. The issues this testing raises are similar to those raised above for any healthy person, but there are also important differences. One significant difference in prenatal screening compared with newborn, child, or adult genetic screening is that parents can choose to terminate the pregnancy. As discussed in more depth later, several ethical issues arise: What level of confidence is required for prenatal screening in order to terminate a pregnancy? Is it appropriate to screen for and act upon nondisease conditions, such as height or hair color? Even if the fetus is diagnosed with a disease risk and the parents decide to continue with the pregnancy, the knowledge of the disease risk might affect the manner in which the parents treat or behave around the child after birth. Parents may become overly concerned or overly protective if they believe their child is at risk for particular conditions. In some cases this behavior may prove beneficial, but in other cases, overprotective behavior may be harmful.

Can genetic testing be useful for choosing healthier embryos and producing designer babies?

In addition to prenatal genome sequencing, it is now possible to analyze the DNA of early embryos, before the embryo even implants into the mother's uterus. Specifically, after in vitro fertilization (IVF), a small number of cells (~1–6) can be removed from early embryos and analyzed by genetic tests. This procedure can be performed by biopsying an 8-cell embryo at the third day after fertilization or the trophectoderm of a blastocyst at the fifth day. In the United States, embryo biopsy typically occurs at the fifth day. Typically, only one or a handful of genes are assayed.

Presently, genetic testing of embryos is only performed in several circumstances. In cases in which a child is born with a severe genetic disease, parents often turn to IVF and embryo testing to avoid having a second child with the same disease. Similarly, if the parents are genetic carriers of a particular disease and they want to ensure that their child is not born with the genetic disease, they may chose IVF screening. IVF is often used to help solve infertility problems, and such embryos can be screened for aneuploidy. Indeed, although some aneuploid embryos can develop and result in live births, most aneuploid embryos are not implanted and many would naturally miscarry. By screening for embryos with the normal numbers of chromosomes, the success rate of IVF can be increased, along with the likelihood of having children without genetic diseases caused by aneuploidy.

Because it is now possible to determine the genome sequence of DNA from one or a few cells (albeit not yet with perfect accuracy or completeness), we can now determine the entire genome sequence of IVF embryos before they are implanted. Could we see a world in which most babies are born after screening in vitro fertilized embryos and those lacking obvious disease-causing variants are implanted into the mother? It is certainly possible and perhaps even likely. This ability to select genetically screened progeny also raises issues about selecting characteristics that are not associated with human disease, such as eye color, height, athletic ability, intelligence, and additional traits as they become better understood. These possibilities, which raise ethical concerns about how genetic information will be used, are discussed later.

It is also possible, in principle, to modify embryos by inserting or modifying genetic material. This has been performed in mice or other mammals for research purposes; genes have been inserted to fix genetic defects (e.g., cystic fibrosis in mice) or enhance traits (e.g., to produce transgenic livestock). A "reporter" gene has even been inserted into the embryos of chimpanzees. These technologies raise the prospect that it might be possible to not only select desired embryos for implantation, but to also modify them to correct disease or enhance traits before a child is born.

10

EFFECTS OF THE ENVIRONMENT ON THE GENOME AND EPIGENETICS

How does the environment contribute to health?

In addition to genetics, environmental factors and age dramatically affect a person's risk for particular diseases. Environmental factors include diet, pathogens such as bacteria and viruses, chemical exposures, and stress. Examples are shown in Table 6. All of these together, along with one's genome and physical activity such as exercise, contribute to one's health in a complex and poorly understood fashion.

When did people first begin to study environmental effects?

One of the first large studies to investigate complex diseases was the Framingham study. This project began in 1948 and initially studied a variety of factors for increased risk of stroke and heart disease in approximately 5,000 individuals. By studying many thousands of people for an extended period of time (several decades), this study identified a variety of factors important for people's health. Smoking, obesity, high blood pressure, high cholesterol levels, and other factors were found to be linked to heart disease. These observations led to important recommendations that are still used by physicians

Table 6. Examples of Environmental Contribution to Disease

Environmental Condition	Increased Disease Risk
Human Papilloma Virus (HPV) infection	Cervical and throat cancer
Infection by unknown viruses	Type 1 and perhaps type 2 diabetes
Hepatitis C infection	Liver cancer
Many chemicals	Cancer
Smoking, air pollution	Many diseases: for example, cancer, atherosclerosis, diabetes
Stress	Type 2 diabetes

for their patients. Later, analysis of DNA from this group and many other large-scale studies identified many genetic regions associated with heart disease, obesity, and diabetes.

When do environmental effects begin?

Much of your growth occurs in your mother's womb and, not surprisingly, at least some environmental effects occur during fetal growth. Examples include the administration of thalidomide during the 1950s to prevent nausea during pregnancy. Many children of women who took this drug were born with abnormal or missing limbs. (Interestingly, the drug has been resurrected in recent years to treat multiple myeloma, a cancer of bone marrow cells.) Exposure of pregnant women to thalidomide and serious viral infections have been shown to be associated with an increase in autism. Interestingly, it has been noted that the frequency of autism in the United States has risen sixtyfold in the past thirty years. Although some of this increase may be related to more frequent diagnosis, the high incidence (estimated currently to be as high as 1 in 45 births) and evidence that genetic factors cannot explain all cases, strongly suggests that environmental factor(s) may be involved.

Another potential source of effects during prenatal development is maternal nutrition. It has been noted in several studies that children and grandchildren born to parents who have conceived during times of food scarcity have smaller children on average. A recent example has shown that Gambian children conceived during times of less food (the rainy season) have lower birth weights relative to those conceived during times of more plentiful food. Similarly, babies born to heavier mothers have been reported to be healthier relative to thin mothers in underdeveloped regions of the world.

In addition to the prenatal period, environmental factors likely affect our health for the remainder of our lives. As noted earlier, chemical exposure, exercise, diet, stress, and pathogen exposures can all impact a person's health regardless of age.

Can environmental factors directly impact the genome?

The environment can directly affect one's DNA. Examples include ultraviolet (UV) radiation from the sun; this causes bases in your DNA to become crosslinked and ultimately causes genetic mutations, which can result in skin cancer. Other examples include smoking (which can lead to cancer in the lungs, esophagus, pancreas, etc.) and exposure to chemicals such as alkylating agents (used, for example, in treatment of textiles) as well as naturally produced compounds (e.g., aflatoxin B1). All of these can directly modify one's DNA and lead to genetic mutations. If these mutations affect oncogenes and/or tumor suppressor genes, they can cause cancer.

What is epigenetics?

Another way that environmental factors impact the genome is indirectly, through epigenetic ("beside the genome") changes. Epigenetic changes refer to stable, heritable changes in the expression of genes that occur without alterations to the actual DNA sequence. Epigenetic changes are important in normal human development and health. For example, these changes in aggregate, that is, the epigenome, contribute to the process

by which the tissues in one's body develop different characteristics although all are composed of cells that share the same
genome. Epigenetic changes also have an important role in a
person's physiologic response to the environment (discussed
in detail later). It is likely that epigenetic changes have an
equal or even greater effect on our health than genetic changes.

Two common types of epigenetic changes are DNA methylation and chromatin modification (Figure 21). DNA methylation
is a process by which one of the nucleotide bases, cytosine, is
directly modified by an enzyme that transfers a methyl group
onto the base. Cytosine methylation often occurs in the upstream
regulatory regions of genes (i.e., the promoters) and usually
results in gene inactivation. DNA methylation also can occur at
other locations outside of promoters. Chromatin modification is
another common type of epigenetic modification. Chromatin is
genomic DNA packaged with proteins. These proteins, called

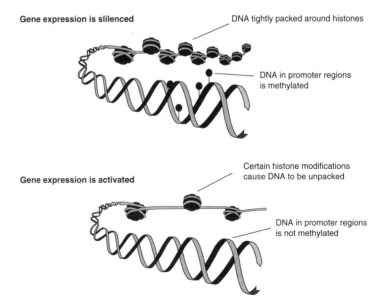

Figure 21. Both DNA methylation and histone modification can affect gene regulation. DNA
methylation in promoter regions generally turns genes off. Modifications on histone proteins can
either activate or inactivate gene expression, depending upon the particular modification.

histones, can be modified in various ways, such as by methylation or acetylation. Simply put, these modifications affect whether the DNA regions are accessible (euchromatin) and the genes therein are expressed; alternatively, the DNA may be packaged in a closed conformation (heterochromatin) and the genes lie dormant. DNA accessibility changes within cells throughout organismal development to give rise to cells of different tissues.

What are some examples of environmental effects on physiology that are mediated through epigenetics?

Both the environment and age can directly or indirectly affect DNA methylation and/or chromatin modifications in ways that have yet to be fully understood. Parental nutritional status has been associated with DNA methylation at specific loci in their children. Returning to the example of season-dependent birth weight of Gambian children, several gene promoters were found to be differentially methylated in children conceived during the rainy or dry seasons. It is hypothesized that these differences cause changes in gene expression, which eventuate in different birth weights. Differences that emerge between identical twins (who share the same genomic DNA) in physical features and health outcomes as they get older are believed to be due in part to epigenetic "drift," the process by which different environmental exposures result in different epigenetic changes.

Exercise has long been known to have an effect on health, lowering the incidence of cardiovascular disease, obesity, and type 2 diabetes. Recently, it has been found that exercise can have a direct effect on DNA methylation and gene expression. A group of 23 subjects exercised one leg but not the other for 45 minutes four times a week for three months. At the end of this period, the methylation and gene expression patterns were different in the exercised leg relative to the one that did not exercise, indicating a direct effect of exercise on epigenetics.

Aging is also associated with specific changes in our DNA methylation patterns. Research suggests that it is possible to predict chronological age within a five-year window based on the methylation pattern in our cells. This has possible applications for forensic analysis—biological samples from a crime scene could be used to estimate the chronological age of the person from whom the tissue came.

It is important to note that many of these links between epigenetic modifications and conditions are *associations* and not necessarily *causations*, that is, it is not always clear what is cause and what is effect. Additional research, often in model organisms, will be important to fully understand the biological effects of epigenetic modifications.

What is the role of epigenetics in disease?

The role of epigenetics in disease is an area of active research. Epigenetic changes have been detected in conditions as diverse as depression, autoimmune diseases, asthma and chronic pain. Both DNA methylation and histone modifications have been implicated in the development and progression of cancer. The enzymes responsible for these modifications are frequently mutated in certain kinds of cancers, and these modifications at specific genes are often altered in tumor cells. As an example of the latter, the hypermethylation of the *RB1* gene promoter was found to decrease markedly the expression of this tumor suppressor, leading to the pediatric cancer of the retina known as retinoblastoma. Further evidence of the importance of epigenetics in cancer biology has been the approval of agents that modify the epigenetic state of cells for the treatment of certain blood cancers and disorders, namely, inhibitors of histone deacetylase and DNA methyltransferase.

How will increased understanding of epigenetics impact health care?

Epigenetics is expected to have an increasingly important role in disease diagnosis and treatment, especially in cancer. Identifying methylation or its readout (i.e., altered gene expression) at cancer-causing genes such as *BRCA1* will likely be a useful diagnostic and possibly prognostic marker (Figure 22). Indeed, *BRCA1* methylation is associated with shorter survival than *BRCA1* mutation. Similarly, detecting changes in chromatin modifications is expected to have utility.

In addition, there are instances in which DNA methylation status may be used to infer drug sensitivity and guide treatment. For example, DNA methylation status has value in predicting response to alkylating chemotherapy (temozolomide) in glioma. Alkylating chemotherapy works by damaging DNA and killing tumor cells. The level of expression of the DNA repair enzyme, MGMT, influences the effectiveness of alkylating chemotherapy—low levels of MGMT correlate with increased chemosensitivity. Methylation of the promoter of the *MGMT* gene (which represses gene expression) is correlated with better responses to temozolomide in glioma.

Eventually, it is likely that screening for disease risk in healthy people will include analyses of both the genome and the epigenome. (See Figure 23 for an example from the author's

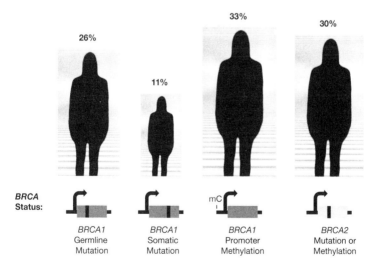

Figure 22. DNA methylation and *BRCA*. The distribution of ovarian cancer genomes with alterations in their *BRCA1* and *BRCA2* genes based on mutations and DNA methylation. Although the sample size was small (103), the analysis clearly revealed that *BRCA1* can be inactivated by either mutation or DNA methylation. *BRCA1* methylation is associated with shorter survival than *BRCA1* mutation.

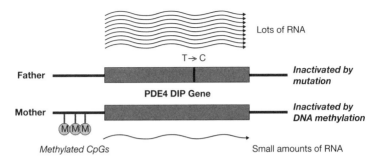

Figure 23. Analysis of the genome and epigenome reveals inactivation of *PDE4* by a combination of deleterious mutation in one copy of the gene and methylation of promoter DNA in the other copy of the gene. Expression analysis showed that only the gene with the deleterious mutation was highly expressed, indicating that both copies of *PDE4* are not functional.

genome.) Genes suspected to be aberrantly expressed—either due to DNA methylation, mutation, or both—will be identified. Findings will be confirmed by gene expression analyses. Thus, in the future we can expect that genomic and epigenomic analyses will both be utilized for predicting disease risk and integrated into health management strategies.

11

OTHER 'OMES

What other 'omes are useful for medicine?

Although the genome and epigenome can provide substantial information about our health, they do not give a complete picture of a disease and how it should best be treated. Other detailed molecular "omic" analyses also can provide a considerable amount of valuable information. The transcriptome, proteome, metabolome, and microbiome have all proven valuable for understanding and sometimes preventing and treating disease. The transcriptome is the collection of RNA expressed from our genes; the proteome is the collection of proteins made from the RNA; the metabolome is the collection of metabolites that we ingest or that are produced by our bodies and the microbes that live in our gut. The aggregate of these microbes is termed the microbiome (discussed later). Because gene expression is the output of both genetic and epigenetic alterations, measurement of the other 'omes provides a means of following their effects. Each of these various layers of the underlying processes of cells can provide valuable information about our health and identify biomarkers and biological pathways that are active or dormant during states of disease or health.

The complexity of each 'ome increases with its downstream product. The average gene makes at least seven different RNAs that often encode different proteins. The proteins themselves can be modified to other forms, which can have differing levels of activity. The different RNA isoforms and protein modifications can give valuable information about the bio-chemical pathways that are active. In addition, the numerous compounds profiled by metabolomics can provide information about the endpoints of gene function and protein expression.

How can the transcriptome and proteome be useful?

Because each type of molecule reflects the activity of cells and tissues, these molecules are thought to be excellent indicators of health states and prognostic outcomes. Indeed, sequencing of transcriptomes of cancer tissues has revealed novel RNA isoforms that are differentially expressed relative to normal tissues. Importantly, the RNA information can shed light on the various biological pathways that are active, classify the tumor into subtypes, and guide treatment. As a recent example, genomic and transcriptomic analysis has stratified colon cancer into subtypes with differing clinical outcomes. Analysis of the proteome of the same tissues reveals an additional colon cancer subtype. Importantly, only about one-third of potential cancer-driving mutations are expressed as RNA; the expressed protein products or associated pathways are logical targets for drugs.

Another example in which the transcriptome has been shown to be clinically useful is prostate cancer. Many cases of prostate cancer are identified early and not associated with any symptoms, are slow-growing, and, depending on a man's lifespan, may not ever cause health problems. It is often difficult, however, to distinguish slow-growing forms of prostate cancer from aggressive forms, so by Stage II many men with prostate cancer have their prostates surgically removed and/or receive radiation therapy. These interventions can have side effects, however, such as impotence or urinary incontinence. Recently, tests have become available that assess the expression of a panel of marker genes in a biopsy sample and can distinguish slow-growing from aggressive forms of prostate cancer, thus helping distinguish between patients who need invasive treatments and those who can have their disease managed simply by regular monitoring or "active surveillance." These gene expression tests are currently performed on a small fraction of the men diagnosed with prostate cancer each year, but this is expected to increase as major medical organizations have begun to incorporate these tests into their prostate management guidelines and large insurers have begun covering them. Similar gene expression assays have a well-established role in managing early breast cancer.

How can the metabolome be useful?

One very important area is the metabolome, which is the collection of all metabolites. These are both synthesized by our bodies and obtained from our food or other sources (e.g., our microbiome, discussed later). Although clearly very important for virtually all human diseases, the metabolome is the least studied of the various 'omes, because, as of yet, no technology has been devised to capture the totality of its complexity in one single assay. Moreover, for the metabolite signatures captured using current methods, only less than half of them can be precisely matched to the existing databases. We do not even know how large a person's metabolome is, though estimates have ranged from a few thousand to a few tens of thousands of molecules.

The metabolome is closely linked to many diseases. In cancer, for instance, the metabolism of solid tumors differs from that of normal tissues in that there is a much higher reliance on glycolysis, which produces lactate, instead of aerobic respiration for energy generation. Indeed, many cancer mutations fall into metabolic enzymes, which affect energy metabolism. For example, the Krebs cycle genes fumarate hydratase (FH) and succinate dehydrogenase (SDH) are mutated in a several cancers as are the *IDH1* and *IDH2* genes, which affect both metabolism and DNA methylation status and are mutated in many different cancers.

Thus far, the study of metabolomics in disease is in its infancy and of the disease associations that have been made, most involve only individual metabolites. For example, high iron is associated with hemochromatosis and *HHR* mutations. Similarly, low levels of folic acid have been associated with birth defects (neural tube malformations); consequently, it is recommended that women take folic acid supplements to reduce the chances of having children with these defects. One exception to the study of individual metabolites is type 2 diabetes, in which both elevated levels of several branched chain amino acids and amino acyl carnitines have been reported. These changes often happen gradually from a low-grade insulin-resistance state to the severe case of diabetes ketoacidosis. Thus, it is likely that identification of metabolites as biomarkers, especially a panel of compounds, could significantly assist early diagnosis and prevention of both congenital and age-related chronic diseases.

Our understanding of metabolomics, personal differences, and human disease will likely evolve as the study of metabolomics becomes more widespread. In particular, it is well known that people metabolize different molecules differently. For example, processing of folic acid can vary up to fivefold among different people, suggesting that women may need very different levels of folic acid supplementation during pregnancy. Furthermore, because many of the enzymes that metabolize drugs and other environmental compounds normally process metabolites, it is likely that our metabolites and requirements will vary greatly among different people. This is already well known for alcohol metabolism, in which genetic variants in key genes (e.g., *ADH* genes involved in alcohol breakdown) affect differences in alcohol metabolism and its effects on people. As our understanding of what constitutes a healthy metabolism materializes, it is easy to envision a future in which people will have their food intake "personalized" through food recommendations and supplements to enhance their health. Furthermore, a better understanding and application of the drug metabolism in individuals will help improve clinical outcome and optimal drug dosing.

How deeply can a person be analyzed?

Studies have been initiated to analyze people as deeply as possible to follow how their biochemical composition dynamically changes during periods of health and disease. The first of these studies analyzed the author very deeply by following his transcriptome, proteome, metabolome, cytokines, antibody reactivity, and more recently their DNA methylation patterns and microbiome, resulting in billions of measurements (Figure 24). These studies revealed the dramatic changes that occur during healthy and disease periods and the types of biochemical pathways that change. By extending these "personal omics profiles" to the study of others it is now clear that each of us

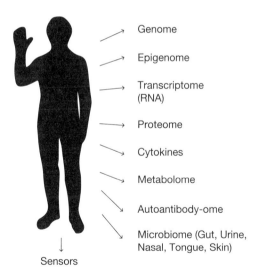

Genome

Epigenome

Transcriptome (RNA)

Proteome

Cytokines

Metabolome

Autoantibody-ome

Microbiome (Gut, Urine, Nasal, Tongue, Skin)

Sensors

Figure 24. Personal omics profiling. It is now possible to analyze people very deeply and make billions of measurements of the different biomolecules present in their blood and urine as well as their microbiome. Activity and physiology can be measured using wearable devices (sensors). By following these over time, we can observe the detailed changes that occur when people acquire disease.

has a unique biochemical profile, and the manner in which people respond to environmental situations can be quite distinct. Such information will help guide people's health by predicting disease and catching adverse events early and on an individual basis.

12

THE PERSONAL MICROBIOME

What is the microbiome?

Although your body comprises an enormous number of cells (10-30 trillion), there are even more—in fact, five to ten times as many cells in and on you that are not you: they are bacteria, fungi, and other microorganisms. These single-celled organisms and other microbes, including viruses, make up your microbiome (often referred to as your microbiota), the collection of organisms that live in and on your body. The microbiome plays an important role in your everyday health. For example, your gut microbiome contains up to three pounds of different types of bacteria, which include many hundreds to thousands of species. These organisms can play both helpful and harmful roles in health. They are responsible for digesting your food and providing many of your essential nutrients and vitamins. For example, bacteria are responsible for providing many essential amino acids through digestion of food and they synthesize crucial vitamins such as vitamin B12, biotin, and folic acid.

It is estimated that thousands (or more) different bacterial species live all over our body. Although the human genome contains about 20,000 genes that contribute to our health status, the microbes that live within and on us collectively contain a staggering 10 million or more different genes; many of these genes make critical contributions to our health. The exact microbial species differs among the different organs such that the microbes in the mouth, nasal cavity, gut and skin each differ from one another. In fact, even different areas of the mouth and different areas of the skin can have very different microbiomes. The microbiome is inherited during birth and is influenced by our environment, such as the food that we eat, including breast feeding. Thus, the microbiome of babies born by Caesarian section is distinct from that of babies born from vaginal birth: the microbiome of the vaginal baby is closest to the vaginal canal of the mother whereas the microbiome of a Caesarian baby is closest to the mother's skin. The gut microbiome shifts dramatically during infancy and early childhood and gradually stabilizes during adulthood. The microbiome differs among different people such that every person's microbiome is unique, meaning that we each have a personal microbiome (Figure 25). Because the microbiome plays an important role in the production of key metabolites, the personal microbiome can be expected to have a significant effect on each individual's health and well-being.

How is the microbiome studied?

The microbiome of any given tissue comprises hundreds to thousands of species, the majority of which cannot be cultured, that is, grown in isolation in a Petri dish. To decipher the various species, samples are usually collected using a sterile cotton swab or spatula and the total collection of DNA from organisms attached to the swab/spatula is isolated. The general types of bacterial species can be elucidated by analyzing a specialized part of the genome that is involved in making a critical component of the ribosome, which is a key part of our protein synthesis machinery. This critical component is the 16S ribosomal RNA gene (rDNA). The 16S rDNA gene is highly conserved but has several differences that are unique to each of the many types of bacteria. The relevant parts of the 16S rRNA genes are amplified and sequenced and the types and numbers of bacteria present in each complex mixture are deduced from the characteristic 16S rDNA of that group of bacteria. The

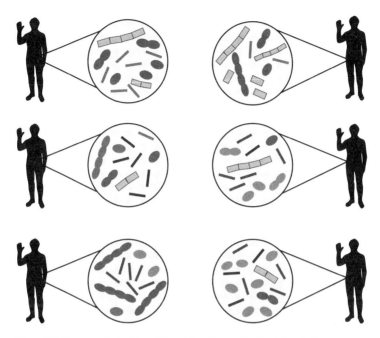

Figure 25. The personal microbiome. We each have thousands of bacteria and other microorganisms living in our gut and all over our bodies. Each person's microbiome is different.

abundance of each bacterial family is derived from the number of times that particular type of 16S rDNA is sequenced.

This type of analysis usually identifies the general families of bacteria and can sometimes reveal the individual species. Although very powerful in answering the question of, "What general types of bacteria there?", this analysis does not answer the questions of, "What exact bacteria are there?" and "What are they doing?" In order to identify individual species and many of the biochemical pathways that are present, many more millions of DNA fragments are sequenced using the latest high throughput DNA sequencing technologies—in fact, the very same ones used to sequence human genomes. These technologies generate a collection of a few hundred base snippets of information across the entire genome of the different bacteria as well as any other organisms in the sample, including viruses, fungi, and other single-celled organisms. The snippets can be assembled into larger segments, and the exact species, their frequencies, and the biochemical pathways they encode are elucidated. The composition of complex samples containing hundreds to thousands of different species can then be deciphered.

How does the microbiome affect health?

The microbiome has been implicated in a variety of diseases. In fact, it has been suggested to have a role in nearly every disease in which it has been studied, including a wide array of metabolic diseases, inflammatory bowel disease, immune-related conditions, cardiovascular and neurological diseases, and cancer. Even depression, anxiety, and autism have been linked to differences in the microbiome. Although the evidence for a direct role of the microbiome has been shown in some diseases, in most cases only an association between bacteria and the disease has been made: it is not clear what is cause and what is effect.

As might be expected given the role of the microbiome in affecting our metabolism, there are clear links between the microbiome and metabolic diseases such as obesity and type 2 diabetes. The gut microbiome of obese people is very different from that of nonobese people. Thinner people have a gut microbiome that often contains a group of bacteria called Bacteroides. In contrast, obese people often have more of another type called Firmicutes. Interestingly, transfer of the gut microbiome from mice that are genetically disposed to become obese into germ-free mice that do not have a genetic predisposition to obesity resulted in increased weight gain and fat accumulation in the germ-free mice. The same results can be reproduced using a human microbiome: Introduction of the microbiome from a thin human into a sterile mouse can prevent obesity, but the microbiome of an obese human cannot. Thus, the microbiome is not only associated with obesity, but it may directly help control weight gain. As with obese individuals, diabetics have a very different microbiome from that of nondiabetics.

A clear role for the microbiome in health has been demonstrated in people with *Clostridium difficile* ("*C. difficile*") infections. These intestinal infections occur in a variety of circumstances including when patients are hospitalized for other illnesses. Transfer of fecal material from a healthy person to one who has *C. difficile* (i.e., a "fecal transplant") has been shown to be incredibly effective in treating this disease. Similarly, people with inflammatory bowel diseases (IBD) such as ulcerative colitis (UC) and Crohn's disease (CD) have problems with their digestive tract. People with CD and UC have a different bacterial microbiome from each other and from that of healthy people. In some cases fecal transplants have caused a dramatic reduction in symptoms. These various observations suggest that the associations between the microbiome and disease may not only be causal associations, but that the microbiome can directly influence health. Moreover, it suggests a novel therapeutic strategy to treat IBD and *C. difficile* patients.

Links also have been made between the gut microbiome and heart disease. Diets high in meat have been linked to heart disease and are comprised of and contain copious amounts of carnitine. Our microbiome converts carnitine into a compound called TMAO, whose levels have been associated with heart failure. Alterations in diet can alter the amount of TMAO, and thus dietary modifications may reduce the severity of heart failure in patients with poorly functioning hearts. These observations indicate that the microbiome can be directly associated with a major human disease and also can suggest a novel therapeutic strategy for reducing heart failure. Given that the bacteria and gene that make TMAO are known, substitution of our indigenous bacteria that make TMAO for one(s) that have been engineered not to produce this compound should improve human health. Although this technology may still be

years away, it offers a much safer prospect to controlling certain diseases using bacterial prophylactics.

How does diet affect the microbiome?

It is clear that alterations in diet have enormous impact on the
gut microbiome. Individuals who switch from a high sugar,
high fat "Western" diet to a low fat or a high fiber diet experience very distinct microbiome changes. Interestingly, a return
to the original diet normally restores a person's microbiome to
its original state suggesting that in general the microbiome is
very stable for a particular person over a long period of time.
Regardless, these results demonstrate the diet can have a dramatic influence on the human microbiome.

Microbial diversity is thought to be important for human
health. A fibrous diet that promotes microbial diversity, in
turn, produces metabolic diversity that is highly beneficial to
our health. Furthermore, the high sugar, high fat Western diets
are thought to lead to a lack of microbial diversity, resulting in
poorer nutritional health. Interestingly, many individuals with
metabolic disorders, such as insulin resistance, are thought
to have difficulty restoring microbial diversity once it is lost.
Thus, prolonged exposure to a Western diet may lead to insulin resistance and diabetes that is difficult to reverse. A better
understanding of how to repopulate a beneficial microbiome
in individuals is desirable and provides an interesting therapeutic opportunity. Overall, it is likely that we will need to
understand our metabolome, microbiome, and food collectively in order to manage our health.

Can the microbiome affect other aspects of our lives?

Recent evidence indicates that the microbiome can affect health
and behavior in other ways. The composition of microbes
on the skin has been associated with whether people are
likely to be bitten by mosquitoes. That is people with certain

natural bacteria on their surface are bitten more frequently than those containing other types of natural bacteria. The exact causal relationship between which microbes attract or repel mosquitoes has not been established. Moreover, microbes are not the only attractant for mosquitoes: they also are attracted to carbon dioxide that we emit from our breath. Nonetheless, it explains why, when outdoors, some individuals seem to be bitten preferentially relative to others. Whether the microbiome affects attraction to other organisms, such as pets, or even whether the microbiome can affect attraction among humans remains to be determined.

Can the microbiome be altered to improve human health?

The microbiome and health can be affected by diet. But what if that is not enough? In principle, the microbiome can be engineered to produce essential vitamins and other useful nutrients, reduce the production of toxic compounds, and even detect and/or remove toxins and adverse compounds. For the latter case, the removal of toxic compounds can be achieved using bacteria that sense the compounds and then induce a detoxifying system. The technologies to be able to manipulate our gut bacteria already exist, but our understanding of the levels of products needed and the possible side effects is unknown and will require considerable testing to complete.

Improving health using engineered bacteria that reside in people could prove difficult. In many societies there is a strong aversion to the use of genetically modified organisms in food—attempts to employ these techniques directly in the human gut could confront a very high regulatory barrier. On the other hand, it likely will be possible to modify the guts of livestock to make foods that are healthier for people: for example, a chicken or pig that produces more essential vitamins. These animals, in turn, could be eaten with minimal consumption of the modified bacteria, which would remain in the animal gut and not be consumed by people.

Finally, although not yet identified, it is also likely that many natural microbes exist in our gut that can remove/metabolize exogenous compounds such as drugs. By identifying and controlling the level of such organisms through diet, probiotics, or perhaps even through genetic engineering, it may be possible to manipulate our ability to respond to medical therapeutics. Overall, it is likely that understanding and manipulating our microbiome will play a key role in managing our health in a wide variety of ways.

13

YOUR IMMUNE SYSTEM AND INFECTIOUS DISEASES

How does your immune system protect you?

You likely have heard of "bubble kids," namely, children who lack an immune system and are forced to live in a sterile environment. If they encounter a pathogen, a virus, or bacteria that can harm them, they have no natural defense and may die.

Your immune system is your defense against all human diseases, including pathogens and cancer. Your immune system is highly complex and comprises over one hundred cell types. These include your B cells, which make antibodies that directly bind to foreign particles, and cells of your innate immune system comprised of T cells and other types, which recognize cells with foreign entities and graft tissue and then attack those cells and tissues. There are "memory cells" in your immune system that can help recognize foreign material that you have encountered previously; upon exposure to the same pathogen, these will quickly stimulate your B cells and T cells to recognize and attack those pathogens and to multiply and defend you. Thus, when vaccinated, for example, with a flu or hepatitis B vaccine, you are exposed to either a dead pathogen or a few key proteins, which will poise your immune system to mobilize and quickly attack the pathogen should it infect you.

Both the antibody-producing B cells and the innate immune cells make a suite of key molecules (antibodies and T-cell receptors, respectively) that recognize the pathogen (Figure 26). These molecules bind the pathogen in the blood, or when it is on the surface of other cells, and help clear the pathogen or infected cells from the body. These molecules form through a process of tuning and selection so that, in the end, they produce highly specific antibodies and receptors. These molecules can be analyzed using a suite of genomic tools, which allow us to follow the exact repertoire of antibodies and T-cell receptors that are produced and define the process.

In order to orchestrate the attack on the immune cells, key stimulators of the immune cells called cytokines are synthesized. These play an important part in regulating the army of cells required to defend against pathogens and adverse cells that the body may encounter. There are large numbers of cytokines, including over 60 that are commonly analyzed.

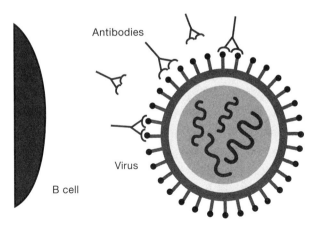

Figure 26. Antibodies are produced by a special type of immune cell called a B cell. B cells attack foreign entities such as viruses and bacterial pathogens.

How does the immune system vary among people and affect our health?

The immune system is the most highly variable component among individuals. This likely reflects the fact that through many generations, different human populations have been preferentially exposed to different pathogens and these populations have developed immune systems that respond accordingly to fight these pathogens. Because of this specialization, there are more genetic polymorphisms in immune-related genes than in any other type of gene in the human body. Examples include the genes that recognize both self and foreign antigens (HLA genes) and the expression of cytokines and immune defense molecules. Moreover, we all have a different history of pathogens to which we have been exposed. As a consequence of both our genetics and environment, we all have a personal immune system poised to deal with environmental and pathogen exposures in a unique fashion. Our ability to fight pathogens varies widely and likely results from inherited and learned differences in our immune systems.

We can now analyze many key players of the immune system at a level that has not been possible previously. From the genetics, we can follow key genes such as HLA to search for propensities for certain diseases. We can also profile the large number of different immune cells and immune molecules in a person's blood. Presently, we can screen for aberrant levels of antibodies (e.g., IgG or IgM) to search for cancer. We also can examine the varying reactivity of different antibodies to different parts of the cell or molecules to screen for autoimmune disease. In the future, it should be possible to take this to a whole new level, whereby analyzing the repertoire of immune molecules for the proteins and pathogens they recognize, and examining the immune system response to challenges in vitro, should give a much more precise understanding of how well prepared our immune system is to fight specific diseases.

How can we analyze infectious diseases?

Most of the time when people become sick, it clears quickly and we say they have been cured. We often do not know how they became sick and how lasting the effects are. For example, people who have a fever rarely know the exact origin of their illness. They rarely even know whether it is a virus or caused by bacteria, or whether there may be long-term consequences to their illness.

Understanding the cause of illness is not often thought to be important, given that most illnesses are transient. In some cases, however, the effects are more lasting. Assays currently exist to diagnose many common viruses, such as rhinoviruses, influenza, adenoviruses, respiratory syncytial viruses, and to identify various types of bacteria, including subspecies that may be particularly pathogenic. The new sequencing technologies not only allow us to identify "all known human viruses" but also the potential for subtypes as well as the identification of new ones. This is important for several reasons. First, we can identify the exact cause early. Diseases such as mononucleosis often are diagnosed only when the illness has refused to go away—in fact, the diagnosis is often made well after others have been infected! Early and precise diagnosis can be extremely valuable for the prevention of epidemics. For example, early detection of serious pathogens such as those causing severe acute respiratory syndrome (SARS), avian influenza ("bird flu"), H1N1, and Ebola virus will be extremely beneficial to the population as a whole. In other cases, detailed analyses can help find the causes of diseases that remain mysterious much longer than they should. A recent case of encephalitis in a critically ill child was solved by sequencing of the child's cerebrospinal fluid; the illness was found to be due to a group of bacteria called Leptospira. Treatment with antibiotics cured the child. In the future, this type of diagnosis should become routine, if not in a primary clinic, then, at least in rapid follow-up clinical testing.

An important benefit of capturing this information is the possibility of determining long-term effects of diseases. There are clear links between serious illness during pregnancy and autism and other diseases, but we do not know which diseases lead to which effects. Type 1 diabetes is believed to be associate with a pathogen infection. Are some viruses more likely than others to cause this disease? If so, which viruses? This information will be be extremely valuable for how we predict risk for particular diseases. For example, an expectant mother with a particular viral or bacterial infection might have her child monitored later for diabetes or autism to enable early intervention.

Finally, another benefit of performing detailed analyses of viruses/bacteria and hosts is to understand their spread. Viruses and bacteria tend to mutate at high frequency and accumulate genetic changes (which is how they often escape the immune system). By analyzing pathogen genomic sequences, it is possible to delineate their spread through their unique changes as well as identify the origin of infection and its pattern of transmission. For example, a recent methicillin-resistant *Staphylococcus aureus* (MRSA) infection outbreak in the United Kingdom was analyzed by genome sequencing of the MRSA strains and subsequently traced to a specific caregiver who had contact with the earliest MRSA-infected patients. This individual could then be isolated to prevent further spread of MRSA. Similarly, on a broader scale, the origin of the recent Ebola outbreak could be traced to its original location, as well as to the likely individual of origin by sequencing the viral genomes and following the pattern of genetic changes. Such information can be useful in evaluating the virulence of pathogens, identifying contacts of those with deadly pathogens, and attempting to quarantine potentially infected individuals.

14

AGING AND HEALTH

The number one risk factor for nearly all human diseases is age. The chances of developing cancer, diabetes, coronary artery disease, heart failure, macular degeneration, and dementia all increase with age. Aging is not a default process in which we "wear out" but rather a regulated process that is controlled by both genetics and environmental factors.

Are there genetic factors underlying longevity?

The ability to live long lives is clearly heritable: children of parents who live exceptionally long, often live long themselves. Indeed centenarians (those who live to be age 100 years or more) can often be found in families. Centenarians avoid most age-related diseases (cardiovascular disease, dementia, cancer) and display very healthy metabolic profiles. Conversely, there are rare diseases that cause premature aging, such as Hutchinson-Gilford progeria, in which mutations present in the *LMNA* gene cause children to age very quickly. These children have wrinkled skin, heart disease, and characteristics of 80-year-olds; nearly all die before age 13. Another premature aging syndrome is the Werner syndrome, in which the gene encoding the WRN helicase is mutated. Werner patients also develop signs of premature aging in their teenage years, with wrinkled skin, gray hair, and cataracts. Both extreme longevity (being a centenarian) and premature aging syndromes have been helpful for understanding the genetic factors that are important for aging. But a key step in pinpointing the genes involved in longevity came from studies in model organisms, where experiments on longevity can be conducted.

Nearly all organisms age. Interestingly, a few species, such as hydra or some species of clams (ocean quahog), show minimal aging. In fact, the latter can live to be over 500 years old (Figure 27)! Many model organisms that can be genetically manipulated, such as yeasts, worms, and flies, have been used to identify and study the genes that control aging. This has

Figure 27. Some species of clams do not seem to age. Unlike humans, the clam depicted at the top is estimated to be 507 years old. Clam image attribution: "Ming clam shell WG061294R" by Alan D. Wanamaker Jr., Jan Heinemeier, James D. Scourse, Christopher A. Richardson, Paul G. Butler, Jón Eiríksson, Karen Luise Knudsen—https://journals.uair.arizona.edu/index.php/radiocarbon/article/view/3222/pdf. Licensed under CC BY 3.0 via Wikimedia Commons.

revealed conserved genes and pathways involved in longevity. For example, the insulin signaling pathway (the insulin receptor itself or the downstream FOXO transcription factors) is critical for the aging process. Interestingly, the insulin-FOXO pathway is conserved across species, and it regulates aging in many organisms, including mammals. Moreover, centenarians have specific genetic variations in genes from the insulin-FOXO pathway. Another pathway that has been shown to be central to aging is the mTOR pathway, which is involved in sensing nutrients, especially amino acids. Blocking the mTOR pathway has been shown to extend lifespan in yeast, worms, flies, and even mammals. Many other metabolic pathways that affect lifespan and healthspan (the portion of life without diseases) have been identified, including Sirtuins (protein deacetylases that are dependent on the metabolite cofactor called NAD+) and AMP-dependent protein kinase (AMPK) (a protein kinase that is dependent on the AMP to ATP ratio in cells). Metformin is a drug commonly used to control type 2 diabetes and an activator of AMPK; it can extend lifespan in mice as well as in people who are diabetic. Metformin has been recently proposed as an "antiaging" drug. Through further study of centenarians and other healthy long lived people, it is likely that more genes that extend human longevity will be uncovered.

Are there environmental factors that affect aging and longevity?

Lifestyle and environment can clearly affect aging. One of the best examples of an environmental condition that delays aging is dietary restriction—a restriction in food intake but without malnutrition. Dietary restriction delays aging and extends lifespan in virtually all organisms tested so far. Although studies on the effect of dietary restriction on maximal lifespan for monkeys are still controversial, even these studies do agree on the fact that dietary restriction delays signs of aging. In humans, dietary restriction is thought to ameliorate metabolic parameters and delay diseases of age, such as cancer. There has been extensive research in the past decade to understand the type of nutrients that need to be reduced as well as the pathways triggered by dietary restriction. It appears that many different regimens can have health benefits, including periods of fasting. Dietary restriction modulates the activity of several pathways involved in aging. For example, dietary restriction inhibits the insulin and mTOR pathways, and activates the AMPK and Sirtuin pathways. Exactly how dietary restriction delays aging is not yet completely understood, but probably involves shifts in metabolism.

On the other hand, certain environmental conditions accelerate aging. Smoking obviously decreases a person's ability to live a long healthy life. Disrupting the circadian rhythms (for example by working night shifts) also has been associated with premature aging and shorter lifespan. There are obvious harmful relationships for people with certain occupations, such as coal miners who develop lung disease. There are also less obvious harmful relationships. For example, nonmanual workers, as well as skilled manual workers (e.g., plumbers, electricians) have longer life expectancies than nonskilled manual workers (e.g., cleaners). However, there are plenty of examples of people subjected to harmful conditions (e.g., smoking) who live long healthy lives. Do such individuals have more proficient DNA repair or other stress resistance systems? The answer is not known.

Does epigenetics control aging?

Interestingly, it has been shown that epigenetic modifications are associated with aging. Gene expression and chromatin modifications (e.g., histone methylation or acetylation, DNA methylation) change with age in many species. Intriguingly, some epigenetic modifications have been proposed to act as "aging clocks." As noted in an earlier chapter, certain DNA methylation patterns correlate very closely with a person's age, suggesting that chemical modifications occur through life. Whether these directly cause aging or age-related features or are simply bystanders of an inherent aging process (like gray hair) is not known. Nonetheless, given that nutrition, exercise, and other environmental effects have been shown to affect DNA methylation, such DNA modifications might accumulate during a person's lifetime and directly influence gene expression, thereby providing a molecular explanation for the aging process. Identifying and studying individuals who are chronologically old, but have the DNA, physical, and intellectual characteristics of a younger person will be very significant for unraveling the mechanism for retarding aging.

Will we someday be able to control our aging?

It is possible that as we learn more about the processes that control aging, we might be able to regulate this process in a fashion that also enables physical and mental longevity. Certainly, maintaining general lifestyle practices, such as exercising and avoiding smoking and overeating will be effective. However, it also seems plausible, if not likely, that using drugs that affect key metabolic pathways such as the mTOR and AMPK pathways, and even manipulating the microbiome through diet and probiotics, might someday be effective retardants of aging. In fact, trials to test metformin, the AMPK activator for reducing aging are presently underway.

15

WEARABLE HEALTH DEVICES

*What other types of personal health information
can be readily collected?*

Most health-related measurements are administered in or
through a doctor's office and are typically taken when we are
sick; the measurements that are taken when we are healthy are
infrequent, and we often do not know what our baseline personal measurements really are when we are healthy. In addition, many types of common events that might influence one's
health, such as pathogen infections, or changes in eating habits
or other lifestyle choices, are often not captured in any system-
atic way, if at all.

New technology is emerging that allows continuous monitoring of physical activities, as well as physiological parameters and biomolecules. Much of this technology consists of
wearable devices (sometimes referred to as sensors or "track-
ers") such as Smart Watches and wrist bands that continuously
record one's movements and exercise (steps, cycling, running),
heart rate, skin temperature, sleep, and stress (often measured
as skin conductance due to perspiration). Many other portable
devices measure important physiological parameters such as
electrocardiograms, oxygenated hemoglobin, fat content and
weight, although not continuously. There are hundreds of such
devices on the market and a partial list is presented in Table 7
and Figure 28. Some of these devices are even free applications that can be downloaded onto a mobile smartphone.
Inexpensive life-logging devices exist that take photographs

Table 7. Examples of monitoring devices. There are now hundreds of wearable devices or "sensors" that follow physiology and activity and biochemical molecules

Parameter	Device/Manufacturer (examples)
Physical activity (Steps, Cycling, Running)	iPhone (Moves); Basis, Fitbit, Jawbone, AppleWatch
Heart rate	Apple Watch, Basis, Scanadu
Skin temperature	Basis, Scanadu
Galvanic skin response	Basis
Sleep quality	Basis, Beddit, (SleepyTime—App)
Oxygenated hemoglobin	Scanadu, iHealth
Blood pressure	Qardio, Scanadu
Glucose	Dexcom
Calories/Nutritional log	MyFitnessPal, Cron-o-meter—Apps
Weight, Body mass index (BMI), Body fat percentage	Withings, QardioBase
Photography of activity (color sensors and passive infrared sensor)	Autographer
Global Positioning System (GPS)	Moves—App
Electrocardiogram	QardioCore, Scanadu, iHealth
Mood	Muse, Emotiv, Melon Headband,
Muscle exertion	Athos, OmSignal
Fat content	Skulpt
Radiation exposure	RadTarge II

at high frequency (e.g., every few seconds) and thereby record your activities at high resolution. There are even such devices for pets. These different devices such as Smart Watches can make over a million measurements on a single person each day.

It is likely that some of these devices, such as the Smart Watches, will become standard equipment that nearly every person will use. These devices will soon have the potential to warn of adverse events well in advance, such as an abnormal heart electrocardiogram predicting a future heart problem.

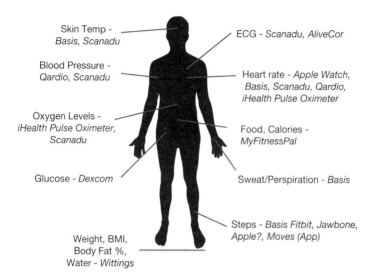

Figure 28. The bionic person. Examples of wearable devices that can be used to continuously or periodically track physiology and activities.

However, even common illnesses might be flagged early. For example, when a prolonged elevated heart rate, increased skin temperature, increased perspiration, and decreased appetite, all occur together, this might suggest the early onset of a pathogen infection before the person is even aware of it.

More recently, new devices have been invented that can continuously measure important biomolecules (also called bioanalytes). There are contact lenses and skin attachments that measure glucose levels in the body, and it is only a matter of time before devices that continuously follow many more bioanalytes become available. In the future, these devices can be personalized such that key molecules important for the individual can be monitored. Add these to the already established invasive devices that control and monitor heart beat, and those that can photograph and analyze our food, and we are gradually progressing toward a bionic world where we will have at our disposal a variety of mechanical devices that continuously measure a multitude of factors important for our health.

How will this information be made available to, and used by, the individual?

Nearly all of these devices can transfer the information directly to a portable smartphone. As such, the smartphone is soon to become the information control system for managing your health status. The information is processed and then can be fed back to the user in a variety of fashions in real-time. These include daily reminders to perform activities, weekly and monthly summaries, positive feedback when personal records (such as miles run) are achieved, and negative feedback when falling below personal milestones. The bioanalyte devices are particularly interesting because they will enable people to see exactly how they react to specific foods. If a banana or other food causes a rapid increase in glucose in a particular person, that can be easily tracked and flagged, and a person can avoid or reduce intake of such food in the future. In this fashion the smartphone will become even more of a valuable accessory and may even control many aspects of our lives.

These sensors and information systems will go well beyond physiologic parameters. Efforts are now underway to develop applications and programs that will be able to read communications (e.g., e-mail) and other activities (Internet browsing) to monitor psychological states such as depression or anxiety, and may help in the early diagnosis of diseases such as bipolar disorder and schizophrenia. Information systems that measure young children may enable early detection of autism and early intervention—such efforts using video recording are already underway. Integration of these behavioral activities and information with other types of genetic and molecular information will be extremely valuable in helping to actively manage personal health. The acquisition of such information will enable medicine to shift from managing and treating disease, to instead predicting a person's disease risk and early detection of events, thereby allowing the individual to stay as healthy as possible prior to full blown disease manifestation.

16

BIG DATA AND MEDICINE

"It is a capital mistake to theorize before one has data. Insensibly one begins to twist facts to suit theories, instead of theories to suit facts."
Arthur Conan Doyle (1891) "A Scandal in Bohemia"

How much data can be gathered about a single person?

Collecting the combined personal omics and other relevant data from a single individual—for example, the genome, transcriptome, proteome, metabolome, microbiome, as well as exercise, biomolecules, and environmental exposure information, plus imaging data from PET scans, MRIs, radiographs, etc.—will lead to enormous volumes of data associated with each individual (Figure 29). How much information is this? A genome sequence normally takes about one half terabyte of space, while other molecular omics information, depending upon the type, can represent several terabytes; thus, with multiple time points collected during a person's life, hundreds of terabytes of data can easily be collected. Imaging data can be much larger and and push this number even higher. Thus, it is plausible that one to several petabytes of information can quickly be generated for a single person. Although it may be possible to reduce the amount of data by discarding raw data and simply saving summaries, this might be a huge mistake, at least in the short-term, as new algorithms are continuously developed that improve the analysis of information that can be extracted from raw data.

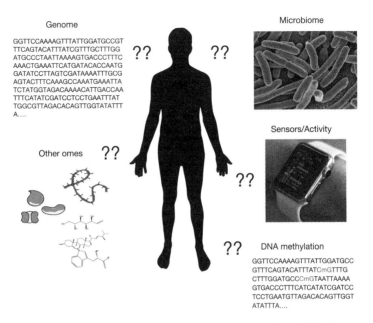

Figure 29. Large amounts of data will be collected about each person. This includes information about each person's genome, microbiome, other omics information, and wearable devices. Both individuals and physicians will need to know how to work with these data.

It seems likely that during the initial development phase of personalized medicine, individuals will get their raw data reanalyzed frequently, resulting in an improvement in interpretation.

How much data can be gathered about a group of people?

We are not far from a world where there will be millions of people with their genomes sequenced and linked to electronic health records. The addition of other types of information—molecular, physiological, physical activity, environmental, etc.—will result in large databases in which genetics, disease outcomes, drug responses, and molecular information will all be linked. This will enable the discovery of new associations of genetic risk with molecular biomarkers and with diseases and drug responses.

The biggest challenge with such databases is collecting the information in a format that is readily sharable. Medical information is extremely heterogeneous and many different terms can refer to the same condition. For example, to measure the ability of blood to clot, a test was developed in the 1930s called prothrombin time. The reagents used to perform this test varied widely, however, and the prothrombin time reported from different labs could not be compared until the international normalized ratio was developed to standardize values across labs. Similarly, even common parameters such as blood pressure can be collected in different ways, such as through noninvasive cuffs or through invasive arterial catheters, and these differences may result in different values. Thus, it will be important to bring data into a common standardized format with as much information as possible about how the measurements were made so that meaningful comparisons can be made.

Another challenge with such databases is that they will likely contain a great deal of private information, and thus the information may be difficult to share due to privacy concerns (discussed later). It is likely that rather than aggregate all medical data into a few large databases, it will be important to interrogate individual data storage sites with software algorithms in order to be able to extract useful information.

How can a large database assist in medical care?

For most diseases doctors prescribe therapies based on their experience and best guess. They do not have precise information about what treatments lead to what outcomes—standard treatment guidelines exist for only a limited number of diseases.

The advantage of having large sums of shared data available leads to an entirely new medical paradigm. Information can be accessed in real time and used to manage an individual's health in a much more precise and quantitative fashion

(Figure 30). If a particular person becomes ill with a particular disease, for instance, a cancer associated with a particular genetic mutation and molecular profile, then in principle, algorithms can and should exist that search all other patients with the same disease and mutation (and ideally a complete genetic and molecular characterization), how they were treated, and what the outcome was. In this fashion treatments can be driven by data, not hunches, thus resulting in a new era of "data-driven medicine." Moreover, each individual who is treated will exhibit a specific response, which can be added to the "living database," thereby providing information that will improve accuracy for future patients. This process requires a

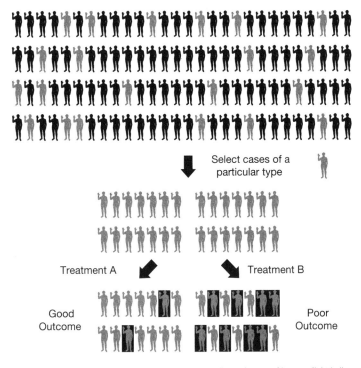

Figure 30. Databases can be searched to find cases similar to the one of interest (light individuals). By examining how they are treated and the outcome (poor outcome outlined in black) a physician can determine the best mode of treatment.

common format and sharing of information, topics that are discussed in more detail later.

It is important to note that this model of a "living database," in which observational data is collected and utilized clinically, is very different from the manner in which drugs are currently vetted for clinical use (Figure 31). Presently, randomized trials are run in which one set of patients are provided a drug and the other set are provided a placebo; the results are collected and analyzed after a certain time span. These trials are expensive and slow, generally costing hundreds of millions of dollars and taking years to perform. In contrast, collecting pooled data that already exists can be much less expensive and can be performed in a matter of days or weeks. It also can be much

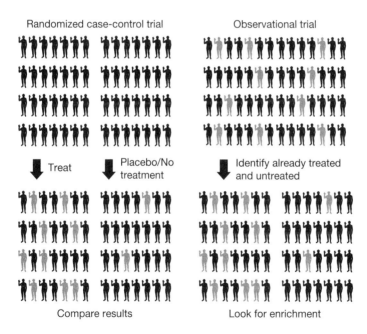

Randomized case-control trial

Observational trial

Treat

Placebo/No treatment

Identify already treated and untreated

Compare results

Look for enrichment

Figure 31. Treatments are currently evaluated in randomized trials in which treatments are given to some patients and not others. This is a slow and costly process. By searching databases for cases similar to the one of interest, outcomes of people receiving treatments can be rapidly deduced and clinical decisions made based on existing evidence.

more precise; for example, it can focus on individuals of the same ethnic or genetic background and in the same geographical locations. One needs to correct for any potential biases, but overall the concept of patients also contributing research information, which, in turn, helps the next patient, is paradigm shifting and will transform the manner in which medicine is practiced.

How can Big Data guide lifestyle decisions?

In addition to direct treatment of disease as described above, Big Data information can be used to manage one's health prior to the onset of disease. This has the potential to shift medicine from a reactive practice of treating symptoms and diseases, to one where disease risk is predicted and managed prior to onset, or is at least diagnosed early. As more and more associations are made between genetic changes and human disease, this information will become commonplace. For example, individuals with mutations in the breast cancer genes will be screened at higher frequency or undergo surgery (which is already happening), while those with alterations in MODY and cardiomyopathy genes will be monitored for insulin and heart defects, respectively, and managed accordingly. As environmental effects become better understood, they will affect one's lifestyle and perhaps even occupation. For example, Parkinson's disease has been linked to pesticides; thus, people at risk for this disease based on their DNA sequence or other markers should probably not be employed where pesticide exposure occurs, and perhaps should use them less frequently in their landscaping.

What are the opportunities for industry in Big Data Medicine?

Industries that can manage large data and extract meaningful information from massive amounts of data will gain enormous opportunities. These efforts will involve data scientists who can handle, integrate, and visualize complex health data, as well as software developers and engineers who can develop new health applications. Large software and information companies such as Apple, Google, Intel, Oracle, Microsoft, SAP and Amazon are already investing heavily in these areas.

17

DELIVERY OF GENOMIC INFORMATION

"As more genomic applications become relevant to patient care, inherent tensions arise between the status quo in health care practice and potentially transformative approaches."
Feero, *Journal of the American Medical Association*, 2013

Who controls your genomic and other health information?

Who controls your genome sequence and other health information? What should be delivered and how? These are important issues. The answer to who controls your genome information should be obvious—you do. Your DNA sequence is yours and you have a right to it. However, the implementation of this concept is not as straightforward as one might think. What information you want and how it is returned requires serious thought. Do you want all of your information returned to you or just the actionable information? What is your definition of actionable? What if there is uncertainty about the accuracy of the prediction?

As an example of what many people would consider nonactionable information, do you want to know whether you have the Huntington's high-risk change? If you have the high-risk change for this gene, your chance of getting this progressive neurological (and ultimately fatal) disease is extremely high. There is no cure, however, so the information will not help you medically; that is, there are no treatments or other lifestyle "actions" that one can take to avoid or lessen the ravages of this disease. Some might argue, however, that this information is still actionable, because you might choose to base your life decisions on this information. Examples of actionable information include the breast cancer risk changes (e.g., *BRCA1* and *BRCA2*) or the risk allele for hemochromatosis; individuals with these mutations should get monitored for disease because there are known treatments to mitigate or even cure these conditions. Ultimately, the decision to receive any genetic and medically relevant information should belong to the patient.

Who will deliver genomic information to you?

Because people may not realize exactly what information they will get from their genome sequence, genetic counselors or people with similar expertise are usually employed to talk to the people receiving the information. Ideally, the potential subjects are educated at the outset about what it means to have their genome sequenced, and then again just prior to when the results are returned to them, in case new questions have arisen. This way, they will have had time to think about the implications of what might be discovered and what it would mean to their families.

Regarding their own personal genome results, possible scenarios include learning that (1) they are likely to be at high risk for certain diseases that may or may not be actionable; (2) they are likely to be a carrier for certain genetic diseases in which one copy of a gene is mutated and the other is fine, and thus, their children could be at risk for a disease; and (3) their ancestry or parentage is not what they thought. They should also be told that (1) genomic sequencing is not 100% accurate; (2) different types of results may be reported by different genomic sequence providers, some of whom are willing to report only exact matches to established disease-causing variants, whereas others report every variant that might be predicted to be damaging, even if not yet a known disease-causing variant. Often, other specific information is returned based on the medical or family history of the subject.

As a general rule, the author believes that it is useful to talk to individuals before they receive their report, in order to try to understand exactly what type(s) of information they want, and then tailor the information in the report accordingly. Although some experts argue that subjects often do not understand the consequences of receiving this information and that it may prove to be harmful in ways the subject did not expect, many people (including the author) believe that the fundamental decision lies with the subject—not the counselor, not the physician, not the government. Moreover, if subjects are entitled to know their family history and results of other medical tests, why should genome sequencing be any different?

What is the role of the physician?

Genome sequencing and interpretation is well beyond the expertise of most physicians, and many physicians are unlikely to prescribe what they do not understand. In the future, physicians will need to receive some general training in this area to understand the basics (see Chapter 20). Moreover, they are likely to encounter people who have had their genomes sequenced, and they should know how to respond if confronted with a question from them. Even if a physician has such expertise, he or she is unlikely to have the time to give any meaningful interpretation of genome sequence information in a typical medical exam visit.

Realistically, as with any specialty area, such as imaging, an expert will need to review the results and formulate a report. Ideally, the expert would have information about the family history and medical condition of the subject, and work with a physician to provide the best possible interpretation and feedback. Importantly, they can discuss the follow-up tests that might be pursued, and additional specialists who might be consulted based on the genomic results. For example, if a new variant is identified that could be suggestive of a hypertrophic cardiomyopathy defect, this can be pursued further with a cardiologist. In this setting the physician helps coordinate rather than dictate the health care of the subject; this happens already, but will become even more prevalent with the mainstream use of genomic information and the resulting early detection of disease risks.

It is likely that in the future reports will become very structured, and physicians will treat the results like any other medical test. One difference with genomic reports of a typical healthy person is that there tends to be a broad range of findings (drug sensitivities, high-risk variants, carrier variants) and with proper training physicians may be able handle much of this. This will likely require a considerable alteration in the way we teach and train physicians. It is also very likely that in the future all genetic counselors will receive intensive training in genomics and become "genome" counselors.

What are the implications of direct-to-consumer genomic testing?

If patients are entitled to their own DNA, shouldn't they have easy and direct access to the data? For the reasons stated above, it is important that the information be delivered in a responsible fashion. Many experts feel that this can only be accomplished in live consultation with an expert. It seems plausible, however, to educate people with an online mini-course, similar to the training courses required by many institutions regarding important topics such as sexual harassment or working with human subjects. For such a "genomic test interpretation" training course, there likely would be the educational content, along with quizzes to assess competence and a questionnaire to gauge the level of information that the subject desires to be reported back.

What if an individual cheats and has someone else take the training course in their place? The answer is obvious—such individuals assume the consequences of the information that they receive, even if it is not what they wanted to hear. And with newer privacy efforts, such as biometric scans for logging into computers or websites, it may be possible in the near future to make such cheating more difficult.

18

ETHICS

Can your genetic information be used against you?

In the United States, a relatively recent law called GINA, for Genetic Information Nondiscrimination Act, prevents discrimination on the basis of genetic information for health insurance and employment. This law has limitations, however, and does not protect individuals from discrimination based on their DNA sequence with regards to life and long-term disability insurance. Most other advanced countries have a minimum level of socialized health care and thus have guaranteed access to many basic services. In such countries, discrimination is less of a concern, at least from a financial perspective.

Whether genetic discrimination in health care, employment, or social arenas actually occurs in practice remains to be seen— the field is young and relatively few people have had genetic testing. We are not aware that any major acts of discrimination have occurred strictly based on genome sequence. It is possible, however, that people with high-risk alleles for certain diseases could be viewed and treated differently. For example, knowing that someone is at risk for Alzheimer's might cause people to assume that they are more forgetful, especially as they get older. Although such circumstances will likely arise, it is worth noting that (a) *everyone* is at risk for at least some diseases— the perfect genome does not exist; and (b) people provide their family histories as part of routine medical care, and this type of information is already available for use by insurance companies. Thus, we could all be subjected to some discrimination.

Unfortunately, personal discrimination presently occurs on the basis of race, ethnicity, gender, sexual orientation, etc. Perhaps DNA will be added to the list.

What are the concerns surrounding routine (or even mandated) genetic screening?

The introduction of genome sequencing into health care has many important implications. In the case of prenatal and early diagnosis in newborns, genome sequencing is likely to prove useful for early diagnosis of disease susceptibility and thus will likely augment existing tests performed at birth. Presently, about 30 such tests are routinely performed. A strong case can be made that genome sequencing or at least genetic screening should be performed for all expectant parents for these purposes. If performed at an early stage of pregnancy, this information may—depending on the wishes of the parents and the severity of the disease—lead to an early termination of pregnancy, or afford the chance to become informed and plan ahead for the birth of a child with a genetic disease.

The situation becomes more complex when genetic variants of unknown effect are discovered and the disease risk is not certain. Similarly, there is a concern about the ethics of termination of pregnancy for non-disease associated situations such as sex (i.e., boy or girl), height, intellectual ability, eye and hair color, athleticism, and other non-health related traits. Presently, most of these traits are difficult to predict from a genome sequence, but it is likely that the predictive power will improve in the near future, particularly when we better understand both genetic and epigenetic effects on these traits. This raises concerns about eugenics, namely, that our society may start producing more and more "prescreened" children who are selected to display only a limited set of genetic traits.

One possible concern with producing a homogenous set of children who are devoid of what we believe to be disease-causing variants is that some variants are likely deleterious in some situations and beneficial in others. For example, the mutation that causes sickle cell anemia is thought to help prevent malaria, while cystic fibrosis mutations are thought to help reduce tuberculosis. Individuals with mutations in the *CCR5* gene do not get HIV infection. Thus, what may be perceived as a disease in one context might prove beneficial in another. From a species perspective, eliminating all disease-causing mutations and creating a more homogeneous population will likely make humans more susceptible to pathogens by reducing the natural genetic variation (including unknown variants that confer pathogen resistance) that is currently present in the human population as a whole.

Should parents have a right to their children's "non-immediate" health information (i.e., diseases that do not occur directly at, or shortly after, birth)? Expectant parents who sequence their children in utero or at birth might become overly concerned about potential risks and over-manage their children's lives and health—picture the so-called "helicopter parents." On the other hand, one could argue that having information that is actionable, especially during childhood, is valuable to prevent and/or ameliorate the effects of diseases before onset of symptoms. For example, it is clear that working with children with autism at an early age can be extremely beneficial. Thus, withholding such useful information could be harmful to the well-being of the child.

What is a possible solution?

It is our belief that everyone should get their genome sequenced as early as possible, ideally before birth and only after counseling of the parents. The immediately actionable information (e.g., metabolic disorders that could be fatal in newborns if not treated) should be returned to the parents to help manage health. Other actionable information that could ultimately be valuable to the child (and thus indirectly to the parents!) would be returned at the discretion of the parents.

The children have a right to access their DNA when they become of age, but again only after counseling. Both parents and children may opt not to ever retrieve their sequence and its interpretation. Furthermore, given that our interpretation of genome sequences will likely improve as more people get their genomes sequenced, the release of an improved interpretation to a person later in life will likely be more accurate.

One advantage to this solution is that should a child experience a difficult health situation, his or her genome could be accessed right away and potentially provide one more useful set of information for solving the disease and/or for managing its treatment. For example, if a debilitating disease occurs, it may be possible to quickly search for possible clues in the child's genome sequence to solve the mystery disease, as described in Chapter 5. In many cases this could help avoid a long diagnostic odyssey.

19

EDUCATION

Michael Snyder and his healthy >80-year-old mother (Phyllis Snyder) both carry a mutation previously thought to inevitably lead to aplastic anemia, but neither have this disease (which illustrates the limitations of genomics).

Can we educate people to understand genomic information?

It has been argued that many people will not understand genetics and genomics, and thus, it will prove difficult to convey genomic information to them. Will people with a gene mutation that raises the risk of developing Grave's disease tenfold from 0.1% to 1% really understand that they still have only a one in one hundred chance of getting that disease? We believe that with proper interaction among healthcare workers, genetic counselors, and patients, the general concepts of genetics and genomics can and must be conveyed. People have a right to this information, much as they have a right to bear children. To help convey information, online sessions, as well as face-to-face sessions with counselors, will ensure that the patient receives the proper education about genomic sequencing.

We have found that people are very eager to understand this information. Basic concepts of genetics already are taught in most high school science classes, as well as in some science museums and public forums such as libraries. Students and the public are generally excited to hear this information. We believe that the average person is capable of understanding basic concepts of genetics, and given its foundational nature in relation to our health, it should be taught to every student in high school.

How do we educate physicians to understand genomic information?

Physicians work hard to serve the best interests of their patients. "Do no harm" is the code they live by. Presently there is a strong feeling among many physicians that genetic and genomic information is harmful and thus should not be used in managing the health of healthy people. Moreover, most physicians are not trained to understand this information. Ironically, many of them are very comfortable discussing family history with their patients but are uncomfortable talking about genomics.

The technologies, applications, and value of genetics and genome sequencing must be part of medical school education and continuing medical education, given how rapidly the field is changing. Most importantly, as alluded to above, patients are truly eager for this information. Genetic information companies such as ancestry.com and 23andMe each have well over 900,000 people who already have contributed samples. Many of these people approach their physicians asking for help in interpreting the meaning of their results. It is essential that doctors be familiar with this subject area.

Who else should we educate? What is the role of healthcare providers, insurers, and policy makers?

Physicians are not the only ones who need to be familiar with genomics and personalized medicine. Healthcare providers, insurers, policy makers, and others will need to become familiar in order to best serve their constituents. Each of these groups needs to serve the individual as well as provide service in a cost effective manner. They will need to understand the value of genomic tests for diagnosing and treating disease as well as for maintaining the health of the people they serve. Understanding value and costs will be essential in order to formulate guidelines and make decisions on the particular tests that should be covered for specific health conditions.

20

PRIVACY

Can people be identified solely from their genome sequence?

Identify theft is a significant concern in the financial sector as compromise of the privacy of financial data can create personal havoc. Is this a significant concern for genomic information? Genomic sequencing has the potential to reveal one's ethnic background and many other personal details, including one's actual identity. A relatively recent study revealed that one could pinpoint the geographical roots of Europeans to within 200 kilometers based on a small snippet of their DNA information. In fact, it is possible to directly identify people based on their DNA sequence when correlated with other publically available information. A recent study investigating a number of released genome sequences coupled with other public information available on the Internet showed that the personal identity of some of the individuals whose genomes were sequenced could be discovered. As more sequences become publically available, this number of identifiable genome sequences will increase greatly.

This raises the possibility that people who leave anything in public that contains their DNA, for example, a hair sample or skin material, might also be identifiable, much like what happens with fingerprinting. Indeed, it has been reported that criminals have been identified from traces of such information left behind at crime scenes. It also means that people who have disclosed genetic information to others can have their information matched against existing databases and thus be identified.

Moreover, a person's genetic information can be used to search a disease database (e.g., Alzheimer's) and reveal whether they or a relative are presently in that database. Thus, in many cases, databases may need to be safeguarded for the types of information that are made publically available or for the linkage of specific disease information to individual genome sequences.

Most importantly, before people consent to have their genome sequences contributed to such disease databases, they need to be thoroughly educated about possible ramifications. We have found that most people with diseases, as well as many healthy people, after being counseled about their options and possible ramifications, still do want to share their genomic information to help solve and treat diseases.

Privacy concerns are not limited to genomic information. The ability to identify people using information from other omes (transcriptomes, proteomes, metabolomes, and microbiomes) in which there is a distinct personal signature will also be feasible someday. Such developments could result in a significant invasion of privacy. As with fingerprinting, however, it remains to be seen whether people will take advantage of such information for nefarious purposes.

Will having my genome sequenced affect my family members?

As noted above, given that half one's DNA comes from one's mother and half from one's father, one's genome information will indirectly reveal some, but not all, of the genome information of one's parents. Similarly, one half of one's DNA is shared with one's siblings (on average) and biological children (exactly), but it is not known which half. Thus, one's genome sequence may suggest things to watch out for in siblings and children, but nothing guarantees that they share one's disease risk mutations. An exception is identical twins, who share the same DNA (except for a small number of de novo and somatic mutations), so that the sequencing of one twin will reveal the genetic risk of the other. Along the same line, cousins will share one-eighth of their DNA and will share some of the same mutations, but not as many as they share with siblings or children. Thus, cousins may share some of the same risks, but because they have even less shared DNA, their risk will be different. Generally, when one family member is sequenced, he or she may want to discuss with other close family members (especially parents, siblings, and children) whether they want to know the results and whether follow-up tests should be performed. If the individual plans to release the data to the public, it would be useful to discuss this with other family members as well.

21

PAYING FOR PERSONALIZED MEDICINE

Who pays for genome sequencing in treating disease?

Although it can be easily argued that genome sequencing can be used beneficially, probably the biggest detriment to implementing genomic medicine is: Who pays? Presently, many insurance companies will reimburse for sequencing the tumor genomes of cancer patients, and the genomes of those with diseases of undiagnosed causes, particularly if it can potentially result in cost savings. As noted earlier, many new drugs can cost $100,000 or more. (See Table 8 for several expensive drugs provided by Medicare.) Cancer genome or targeted gene panel sequencing may identify cancer gene mutations that accurately predict response to specific drugs, and this precision targeting of therapy can justify the costs of sequencing. Similarly, many children have diseases with undiagnosed causes, which can result in large diagnostic and treatment costs. Insurance companies will pay for the genome sequencing of the affected child, and sometimes of the parents and other family members, to help find the likely causative mutation. The identification of the involved gene may also identify novel applications of existing drugs for these diseases, and may be used to develop a test to screen other family members and others with similar symptoms.

Because it is still in its infancy, genome sequencing has not been fully demonstrated to be beneficial in most cases, and

Table 8. Medicare spent $19.5 billion for drugs administered by physicians or "outpatients" ("Part B drugs) in 2010. The 10 most expensive drugs per year are shown in this Table and account for 45% of the total costs of Part B drugs. (According to the Government Accountability Office, http://www.medicarenewsgroup.com/news/medicare-faqs/individual-faq?faqId=48ec6c39-e49a-4bb9-a892-19a6abac8237). Another recent expensive drug is Soliris, which is used for treatment of Hepatitis C and can cost over $200,000 per treatment.

	Drug	Treats	Cost
1	Factor viii Recombinant	Hemophilia A	$216,833
2	Remodulin	Pulmonary arterial hypertension	$130,772
3	Ventavis	Pulmonary arterial hypertension	$84,205
4	Primacor	Acute decompensated heart failure	$62,790
5	Erbitux	Cancer	$25,898
6	Dacogen	Myelodysplastic syndrome	$25,858
7	Herceptin	Cancer	$25,797
8	Vidaza	Myelodysplastic syndrome	$22,957
9	Sandostatin LAR Depot	Acromegaly, diarrhea, and flushing caused by cancerous tumors and vasoactive intestinal peptide secreting adenomas	$22,748
10	Velcade	Cancer	$19,667

thus it is not always possible to justify payment from insurance companies. For situations where people have serious health problems that are not related to cancer or undiagnosed childhood diseases, genome sequencing is presently not usually available in the clinical setting. Such individuals will have to pay out-of-pocket or find researchers who are willing to take on such cases.

Who pays for genome sequencing in preventive medicine?

One might think that healthcare providers or insurance companies should pay people to have their genome sequenced while they are healthy—with this information, such individuals are likely to manage their health better. Presently, however, healthy people pay for their own genome sequencing; only in rare exceptions can it be justified as a potential cost savings. In the United States, there is no incentive from the payers' (insurers) perspective to pay for such sequencing even if it results in an overall costs savings in the long-term. One major reason is that individuals in the United States usually obtain their insurance through their employer. Because the individual can change jobs or the employer can change health plans, it may not benefit one insurance company to invest large sums of money to prevent future disease when the insured individual may not be with their plan in the future.

Preventative healthcare genome sequencing is gaining much more traction in places with socialized medicine, notably Canada, Europe, and Japan. Already large projects such as the 100K UK project and others are being launched to test the role of incorporating genome sequencing into health care. The Million Veterans projects (MVP) in the United States and the recently announced Precision Medicine Initiative have the potential to accomplish this as well. These projects will be valuable in testing whether health care quality, outcome, and costs will benefit from genome sequencing technology. It is also possible that healthcare providers can partner with pharmaceutical companies; in exchange for covering the sequencing costs companies can gain access to medical information that can help develop new drug targets or applications of existing drugs. A partnership of this type has been established between the Geisinger Health System and the Regeneron pharmaceutical company, in which over 100,000 individuals will be sequenced.

Until genome sequencing and other omics technologies become widely implemented, medicine will remain a much more reactive profession in many respects—one in which people largely seek medical advice after a disease arises rather than a preventative profession where information is used to guide lifestyle and help early diagnosis.

Will genome sequencing make health care cheaper?

Many drugs cost hundreds of thousands of dollars per year. Wouldn't delivery of the right drugs to the right patients save a lot of money? In some areas, the answer is very likely to be, "Yes." In the cancer area, giving expensive drugs that can cost $100,000 or more for the course of treatment to patients when they are not likely to respond will realize a considerable cost in savings. (See Table 8 for examples of expensive drugs provided by Medicare.) Similarly, catching individuals with heart or coronary artery problems prior to a heart attack or stroke is likely to reduce these adverse events and help those who might otherwise go on long-term support to live a more productive and healthier life. Solving undiagnosed diseases will cut down on many futile tests as well as on considerable anxiety.

In other areas, genome sequencing is less likely to result in an overall cost savings. Most undiagnosed diseases are not solved by genome sequencing and even if they are, this does not lead to therapies and can result in an additional set of follow-up tests. In the case of preventative medicine, although we expect that most people will find value in genome sequencing for managing their health, it will most likely postpone an adverse event, but may not ultimately save funds in the long term. In this respect, we expect that overall genome sequencing will improve the management of people's health by catching disease early and helping give the right drugs to the right people, but it may not necessarily save money for the system overall. In many cases, the real savings will be in the quality of health care.

Will people act on genomic information?

In order for genome sequencing to be useful for preventative medicine it is essential that those people who have been sequenced act on the information. In many cases, such as the *BRCA* mutations, this is likely to occur. But for many other cases, perhaps the majority, action is a significant problem. As an example, obesity is a significant risk factor for many diseases, including myocardial infarction (heart attack), diabetes, and cancer. The solution is obvious—controlled diet and exercise— yet most overweight and obese people do not do either of these. Why would genome sequencing have a better outcome?

It may not help in preventative medicine if people choose to ignore the information. It is possible in a number of cases, however, that people will begin to use medicines before the disease arises, similar to the case of statins and heart problems. Some examples might be the use of diabetes drugs (metformin) and Alzheimer's drugs that could be more effective prior to the onset of disease. It is very likely that genome sequencing will help in early detection and more accurate diagnosis, much like family history does. Thus, it is likely to have some impact—although whether it will be enough to justify the costs for healthy people remains to be seen. As the price of sequencing continues to decrease, however, the cost will likely no longer become an issue and genome sequencing may become a routine part of a person's health care.

22

THE FUTURE OF PERSONALIZED MEDICINE

What other technologies will be prevalent in the personalized medicine space?

The information gathered from genomics will ultimately be coupled with other technologies to further guide and treat human health. For example, it is now possible to construct pluripotent stem cells from any individual. These are cells that can be grown indefinitely, but can also (with certain treatments) be transformed into many cell types. These cells can be used to (1) test predictions of drug sensitivity, side effects, and efficacy, and (2) determine the potential for analysis of disease-related defects using cell-based assays. For example, if individuals are predicted to have potential cardiomyopathies, their differentiated cardiomyocytes can be examined under various stresses to search for defects. Furthermore, known drugs can be tested on personal cell types to see if they can ameliorate the defects. All of this can be performed in a petri dish to determine if a potential defect exists, prior to the onset of the disease.

In the area of reproduction we now have the potential to produce human children through in vitro fertilization and to manipulate the genomes of such embryos by introducing genetic alterations. Although many safety issues still remain with these various technologies, it will be important to guide these technologies for the betterment of human health, while ensuring adherence to ethical guidelines.

What will the future look like?

In the future we can envision a world in which many people have their genome sequenced, likely before birth, and epigenomes and a wealth of other information will be commonly used to predict, diagnose, and treat disease (Figure 32). Perhaps most importantly, however, this information will be used to keep people healthy. By tracking activities, diet, and molecular information, an integrative information system will manage each person's health.

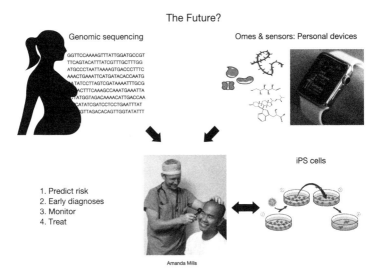

Figure 32. A possible future for personalized medicine. People will have their genome sequences determined before birth. Other detailed measurements will be performed as home tests. Sensors will be used to either continuously or periodically measure our activity and physiology and all of this information will be used to help guide our health care. iPS cells image attribution: "Induction of iPS cells" by Y tambe—Y_tambe's file. Licensed under CC BY-SA 3.0 via Wikimedia Commons—https://commons.wikimedia.org/wiki/File:Induction_of_iPS_cells. svg#/media/File:Induction_of_iPS_cells.svg.

What do these developments mean for the future of medicine? Certainly medical students in training today and physicians in the future need to ensure they are well-educated in genomics and bioinformatics so that they can optimally apply the information generated to patient care. Current medical treatment guidelines based on limited information will be seen as hopelessly antiquated. In the future, treatment algorithms for specific diseases may become entirely computer-based, as comprehensive information about the specific patient and her or his disease (including genomic information) is fed into the system and a customized treatment plan is developed. Although much information will still be managed through a physician, the patient will have much more control. Overall, the information-poor manner in which medicine is practiced today will be replaced with an information-rich system that will be more accurate and more predictive. As patients, we will have more control of our medical destiny and that of our children, and this responsibility will be shifted from the physician to ourselves.

Further Information:

Online personalized medicine course
Stanford Genetics and Genomics Certificate Program
    http://geneticscertificate.stanford.edu/

General
Topol, E. *The Patient Will See You Now*. Basic Books 2015 New York.

Sports
Lippi G1, Longo UG, Maffulli N. Genetics and sports. *Br Med Bull.* 2010;93:27–47. doi: 10.1093/bmb/ldp007.

Pharmacogenomics
Evans WE, McLeod HL. Pharmacogenomics—Drug disposition, drug targets, and side effects. *N Engl J Med.* 2003; 348:538–549. http://www.nejm.org/doi/full/10.1056/NEJMra020526

Cancer and immunotherapy

Brahmer JR, Pardoll DM. Immune checkpoint inhibitors: making
immunotherapy a reality for the treatment of lung cancer. *Cancer
Immunol Res.* 2013; 1;2:85–91.
http://cancerimmunolres.aacrjournals.org/content/1/2/85.full

Useful cancer genomics weblink:

http://cancergenome.nih.gov/cancergenomics

Detailed Omics Profiling

Chen R, Mias GI, Li-Pook-Than J, Jiang L, *et al*. Personal omics
profiling reveals dynamic molecular and medical phenotypes.
Cell 2012;148:1293–307.
PMID: 22424236
http://www.cell.com/cell/abstract/S0092-8674(12)00166-3?_return
URL=http%3A%2F%2Flinkinghub.elsevier.com%2Fretrieve%2Fpii
%2FS0092867412001663%3Fshowall%3Dtrue

INDEX

aging. *See also* longevity
  environmental factors that
    affect, 117–18
  epigenetics and, 118–19
  prospect of controlling, 119
Alzheimer's disease, 64–65
AMP-dependent protein kinase
  (AMPK), 117–19
antibodies, 109–11, 110f
ApoE$_4$, 64–65
athletic performance and injuries,
  using genetic testing to
  predict, 78–79
autism, 9, 59, 61–62, 65, 88, 104,
  124, 141

B cells, 109, 110, 110f
Beery twins, 53–54, 54f
big data, 1
  how it can guide lifestyle
    decisions, 130
  opportunities for industry in
    big data medicine, 130–31
bioanalytes, 123, 124
BRCA$_1$ and *BRCA$_2$* mutations, 25f,
  51t, 72–73
  DNA methylation and,
    92–93, 93f
  how they cause cancer, 24–25

breast cancer, 26
  genes implicated in familial,
    24–25, 25f. *See also BRCA$_1$*
    and *BRCA$_2$* mutations

cancer
  genes implicated in, 24–26, 25f,
    27f, 28t, 35t. *See also BRCA$_1$*
    and *BRCA$_2$* mutations
  genetic causes, 21–26, 23f
  genetics and genomics for early
    detection, 41
  genome sequencing
    deciding whether to get
      one's tumor genome
      sequenced, 37–39
    how it can advance
      treatment, 32–37
    what has been learned, 31–32
  nature of, 21
cancer drugs. *See also* imatinib
  combination therapy, 40
  how DNA can guide drug
    treatments, 67–69
  how genome can guide drug
    treatments, 67
  reasons for failure (and how
    genomic approaches are
    useful), 39–40

cancer drugs (*cont.*)
   sex differences in effects of,
      69–70, 70t
cancer treatment. *See also*
      cancer drugs
   genetics and genomics
      for monitoring the
      effectiveness of, 41–42
   how genetic information
      enhances, 26, 28–29
   how genome sequencing can
      advance, 32–37
carrier mutations, 72
cell proliferation, 21–22, 27f
chorionic villus sampling
      (CVS), 81
chromatin modification, 90–91
chromosomes, X, 12–13
chronic myelogenous leukemia
      (CML), 28–29
circulating tumor DNA (ctDNA), 41
clam species that do not age,
      116, 116f
Clostridium difficile
      (*C. difficile*), 105
copy number variations
      (CNVs), 82
Coumadin (warfarin), 68
Crohn's disease (CD), 105
CTLA-$_4$, 44
cytochrome P450, 67–69
cytokines, 110
cytosine methylation, 90

designer babies, using genetic
      testing for producing,
      84–85
diabetes, 62–63
   type 1, 63
   type 2, 60f, 63, 75, 75f, 88t, 98
      drugs for, 76, 77t, 117
direct-to-consumer genomic
      testing, implications,
      136–37

DNA
   circulating tumor DNA
      (ctDNA), 41
   general features, 5–7, 6f, 8f
   guiding drug treatments, 67–69
   16S rDNA, 102–3
DNA methylation, 90–93, 93f
DNA sequence changes/variants,
      9, 10f, 11–12
DNA sequencing, fetal, 82f, 82–83
dominant and recessive genetic
      mutations, 47, 48f
drug dosage, 67–68
"druggable targets," 29, 34, 35, 37
drug resistance, 39–40
drug response
   sex differences, examples of,
      69–70, 70t
   variants predicting, 76, 77t
drugs, pharmaceutical. *See also*
      cancer drugs
   costs, 151, 152t
   for type 2 diabetes, 76, 77t, 117

embryos, using genetic testing for
      choosing healthier, 84–85
environmental effects, 87, 88t
   on aging, 117–18
   impacting the genome, 89
   mediated through
      epigenetics, 91–92
   when people first began
      studying, 87–88
   when they begin during life, 88–89
epidermal growth factor receptor
      (EGFR), 26, 29, 34,
      34f, 36, 39
epigenetic drift, 91
epigenetics and epigenomics,
      89–91
   and aging, 118–19
   examples of environmental
      effects on physiology
      mediated through, 91–92

how understanding of
    epigenetics will impact
    health care, 92–94
role in disease, 92
estrogen receptor (ER) positive
    breast cancer, 26
exome sequencing, 16, 17f, 18, 31,
    32, 52, 53, 58

fecal transplant, 105
fetal DNA sequencing, 82f, 82–83

gap junction beta-6 protein.
    See GJB6
gender differences
    in cancer drug effects, examples
        of, 69–70, 70t
    in genomes, 12–13
gene fusion, 23
genes, subsections of, 17f
genetically engineered babies.
    See designer babies
genetic diseases
    complex, 61–64
        general prioperites/
            nature, 59–60
        how some diseases can
            be both monogenic and
            complex, 64–65
    how to identify the genes
        responsible, 50, 52–53
    mystery and undiagnosed.
        See also Mendelian diseases
    nature, 47–49
        usefulness of genomic
            approaches to solving,
            53–55
Genetic Information
    Nondiscrimination Act
    (GINA), 139
genetic mutations, 21–23. See also
    genetic variants
    recessive and dominant, 47, 48f

genetic screening, concerns
    around routine/
    mandated, 140–41
genetic variants, 9, 10f, 11–12. See
    also genetic mutations
genome mutations, types,
    72–74, 73f
genomes
    environmental effects on, 89
    functions, 9
    gender differences in, 12–13
    general properties/nature,
        6f, 7–8
    how they are decoded, 13, 14f,
        15–16, 17f, 18
    individual differences, 9,
        10f, 11–12
    strategies for analyzing healthy
        people's, 73f
genome sequences, identifying
    people solely from their/
    privacy, 147–48
genome sequencing, 13, 14f,
        15–17. See also under cancer
    and health care costs, 154–55
    how deeply a person can be
        analyzed, 99–100
    how it can affect the drugs
        someone takes, 76
    how it can improve health,
        71–76
    how it will affect one's children
        and relatives, 79
    impact on one's family
        members, 148–49
    in preventive medicine, who
        pays, 153–54
    in treating disease, who pays,
        151, 152t, 153
    used to identify
        nonchromosomal
        mutations that might cause
        disease, 83–84

genome sequencing technologies and prenatal testing, 81–83, 82f
genomic information. *See also* personal health information
  educating healthcare providers, insurers, and policy makers to understand, 144–45
  educating people to understand, 143–44
  educating physicians to understand, 144
  people acting on, 155
  used against you, 139–40
  who controls your, 133–34
  who will deliver it to you, 134–35
genomic testing, direct-to-consumer
  implications of, 136–37
germline mutations, 22
GJB₂, 52
*GJB*6, 52

health, how genome sequencing can improve, 71–76
health care
  how large databases assist, 127–30, 128f
  how understanding of epigenetics will impact, 92–94
health care costs
  genome sequencing and, 154–55
  who pays for genome sequencing in preventive medicine, 153–54
  who pays for genome sequencing in treating disease, 151, 152t, 153
healthcare providers, educating, 144–45

hearing loss in infants, 50, 52
hemoglobin *β*-globin gene *HBB*, 47–48
histone modification, 90f
human epidermal growth factor receptor 2 (HER₂) positive breast cancer, 26
hypertrophic cardiomyopathy (HCM), 52, 78

imatinib, 28–29
immune system
  how genomics can harness it to fight cancer, 44–45
  how it protects you, 109–10
  how it varies among people and affects health, 111
immunotherapy, 42–44
infectious diseases, how to analyze, 112–13
injuries, sports
  using genetic testing to predict, 78–79
insulin, 117. *See also* diabetes
insurers, educating, 144–45
in vitro fertilization (IVF), 55, 84, 85
induced pluripotent stem cells (iPSCs), 32f

leukemia, chronic myelogenous, 28–29
lifestyle decisions, how big data can guide, 130
"living database," 128–29
longevity
  environmental factors that affect, 117–18
  genetic factors underlying, 115–19
lung cancer, 35t, 41
lymphocytes. *See* B cells; T cells

maturity onset diabetes in the young (MODY), 63
Medicare, money spent on drugs by, 151, 152t
Mendelian diseases, 47
    genes implicated, 49–50, 51t
    general properites/nature, 47–49
    number, 49–50
    usefulness of genomic approaches to solving, 53–55
    why most of them cannot be solved, 55–58
metabolic diseases, 104. *See also* diabetes
    how complex genetics affects, 62–64
metabolome, usefulness of, 97–99
methicillin-resistant *Staphylococcus aureus* (MRSA), 113
MGMT, 93
microbiome/microbiota, 103f
    general properites/nature, 101–2
    how it affects health, 104–6
    how it can be altered to improve health, 107–8
    how it is studied, 102–4
    impact on health and behavior, 106–7
    relation to diet, 106
monitoring devices, 121–23, 122t, 123f
mystery genetic diseases. *See* genetic diseases: mystery; Mendelian diseases

neoantigens, 44–45
neurological diseases, 64–65
    how complex genetics affects, 61–62
non-small-cell lung cancer (NSCLC), 35t, 41

O^6-alkylguanine DNA alkyltransferase. *See* MGMT, 93
obesity, 104
'omes
    how deep can you profile, 99–100, 125–26, 126f
those useful in medicine, 95–96.
oncogenes, 22
1000 Genomes Project, 16

personal factors that impact our health, 2–3, 3f
personal health information.
    See also genomic information
    how large databases assist in medical care, 127–30, 128f
    how much data can be gathered about a group of people, 126–27
    how much data can be gathered about a single person, 125–26, 126f
    that can be readily collected, 121–23
        how it will be made available to and used by the person, 124
    who controls your, 133–34
personalized medicine
    future, 158f, 158–59
    general description, 1–2
    resources for information on, 159–60
personal omics profiles/personal omics profiling, 99f, 99–100
phasing variants, 15f, 15–16
physician(s)
    educating, to understand genomic information, 144
    role of the, 135–36
pluripotent stem cells (iPSCs) p157, f32
policy makers, educating, 144–45

prenatal testing, how genome sequencing technologies are changing, 81–83, 82f
probiotics. *See* microbiome
progesterone receptor (PR) positive breast cancer, 26
prostate cancer, 96–97
prostatic acid phosphatase (PAP), 45
proteome, usefulness of, 96–97
proto-oncogenes, 22

randomized controlled trials (RCTs), 129, 129f
recessive and dominant genetic mutations, 47, 48f
reference genome, 13
RNA, 8, 8f, 95–96

schizophrenia, 9, 59, 61
Segawa's dystonia, 54
sickle cell disease, 47
single nucleotide variants (SNVs), 9, 10f
somatic mutations, 23, 34
sports performance and injuries, using genetic testing to predict, 78–79
SPR gene, 54
16S rDNA, 102–3
Staphylococcus aureus, methicillin-resistant, 113
statin drugs, 68–69

tamoxifen, 68
targeted therapies and approaches, 18, 26, 28–29, 35–40, 52
targeted therapy agents, 35t
T cells, 109, 110
TMAO (trimethylamine N-oxide), 105
transcriptome, usefulness of, 96–97
Turner syndrome, 81

ulcerative colitis (UC), 105
undiagnosed disease See genetic disease: mystery; Mendelian diseases

variants of unknown origin (VUS), 74
viruses, 112, 113
Volker, Nicholas, 53

warfarin (Coumadin), 68
wearable devices, 121–23, 122t, 123f
whole genome sequencing, 18, 52, 53, 58. *See also* genome sequencing